Blockchain-based Smart Grids

Blockchain-based Smart Grids

Edited by

Miadreza Shafie-khah

ACADEMIC PRESS

An imprint of Elsevier

Academic Press is an imprint of Elsevier
125 London Wall, London EC2Y 5AS, United Kingdom
525 B Street, Suite 1650, San Diego, CA 92101, United States
50 Hampshire Street, 5th Floor, Cambridge, MA 02139, United States
The Boulevard, Langford Lane, Kidlington, Oxford OX5 1GB, United Kingdom

Notices
Knowledge and best practice in this field are constantly changing. As new research and experience
broaden our understanding, changes in research methods, professional practices, or medical
treatment may become necessary.

Practitioners and researchers must always rely on their own experience and knowledge in evaluating
and using any information, methods, compounds, or experiments described herein. In using such
information or methods they should be mindful of their own safety and the safety of others,
including parties for whom they have a professional responsibility.

To the fullest extent of the law, neither the Publisher nor the authors, contributors, or editors,
assume any liability for any injury and/or damage to persons or property as a matter of products
liability, negligence or otherwise, or from any use or operation of any methods, products,
instructions, or ideas contained in the material herein.

Library of Congress Cataloging-in-Publication Data
A catalog record for this book is available from the Library of Congress

British Library Cataloguing-in-Publication Data
A catalogue record for this book is available from the British Library

ISBN: 978-0-12-817862-1

For information on all Academic Press publications
visit our website at https://www.elsevier.com/books-and-journals

Publisher: Joe Hayton
Senior Acquisitions Editor: Lisa Reading
Editorial Project Manager: Michelle W. Fisher
Production Project Manager: Sojan P. Pazhayattil
Cover Designer: Matthew Limbert

Typeset by SPi Global, India

Working together
to grow libraries in
developing countries

www.elsevier.com • www.bookaid.org

Contents

Contributors

Numbers in parenthesis indicate the pages on which the author's contributions begin.

Navid Bayati (195), Department of Energy Technology, Aalborg University, Aalborg, Denmark

Rui Castro (5), INESC-ID/IST, University of Lisbon, Lisbon, Portugal

João P.S. Catalão (5, 43, 75, 103), Faculty of Engineering of the University of Porto and INESC TEC, Porto, Portugal

Sijie Chen (169), Department of Electrical Engineering, Shanghai Jiaotong University, Shanghai, China

Juan Manuel Corchado (181), BISITE Research Group, University of Salamanca; Air Institute, IoT Digital Innovation Hub (Spain), Salamanca, Spain; Department of Electronics, Information and Communication, Faculty of Engineering, Osaka Institute of Technology, Osaka, Japan; Pusat Komputeran dan Informatik, Universiti Malaysia Kelantan, Kota Bharu, Kelantan, Malaysia

Ozan Erdinç (103), Yıldız Technical University, Istanbul, Turkey

Ayşe Kübra Erenoğlu (103), Yıldız Technical University, Istanbul, Turkey

Amin Shokri Gazafroudi (181), BISITE Research Group, University of Salamanca, Salamanca, Spain

Matthew Gough (5), Superior Technical Institute, Lisbon; INESC TEC, Porto, Portugal

M. Hadi Amini (61), School of Computing and Information Sciences; Sustainability, Optimization, and Learning for Interdependent Networks Laboratory (Solid Lab), Florida International University, Miami, FL, United States

Amin Hajizadeh (145, 195), Department of Energy Technology, Aalborg University, Esbjerg, Denmark

Seyed Mahdi Hakimi (145), Department of Electrical Engineering, Damavand Branch, Islamic Azad University, Damavand, Iran

Barry Hayes (75), School of Engineering, University College Cork, Cork, Ireland

Hosna Khajeh (75, 131), School of Technology and Innovations, University of Vaasa, Vaasa, Finland

Hannu Laaksonen (75, 131), School of Technology and Innovations, University of Vaasa, Vaasa, Finland

Mohamed Lotfi (43), Faculty of Engineering, University of Porto; INESC TEC, Porto, Portugal

Yeray Mezquita (181), BISITE Research Group, University of Salamanca, Salamanca, Spain

Cláudio Monteiro (43), Faculty of Engineering, University of Porto, Porto, Portugal

Jian Ping (169), Department of Electrical Engineering, Shanghai Jiaotong University, Shanghai, China

Javier Prieto (181), BISITE Research Group, University of Salamanca; Air Institute, IoT Digital Innovation Hub (Spain), Salamanca, Spain

Sérgio F. Santos (5), INESC TEC, Porto, Portugal

İbrahim Şengör (103), Katip Çelebi University, Izmir, Turkey

Miadreza Shafie-khah (1, 5, 43, 75, 131, 181), School of Technology and Innovations, University of Vaasa, Vaasa, Finland

Mohsen Soltani (195), Department of Energy Technology, Aalborg University, Aalborg, Denmark

Saber Talari (75), Fraunhofer Institute for Energy Economics and Energy System Technology, Kassel, Germany

Wei Wei (169), Department of Electrical Engineering, Tsinghua University, Beijing, China

Zheng Yan (169), Department of Electrical Engineering, Shanghai Jiaotong University, Shanghai, China

Chapter 1

Introductory chapter: An overview of the book

Miadreza Shafie-khah

School of Technology and Innovations, University of Vaasa, Vaasa, Finland

These days the advances in communication technologies, IoT, and intelligent smart meters lead the conventional unilateral power grids into intelligent bilateral ones. In newly emerging smart electric grids, distributed renewable resources and demand can play a proactive role, changing the structure of the system from the centralized unilateral scheme to the smart bilateral networks. In conventional grids, power was only transferred from supply side to the consumers that are considered as submissive ratepayers. In a smart grid environment, however, all parties in the system, ranging from customers to the bulk-generating units, can experience real-time communication with each other through state-of-the-art technology. The data from smart meters have the capability to be transmitted in less than a second, leading to better management of the consumption.

Nevertheless, the newly intelligent grids give birth to some challenges, for instance, the possibility of transferring the huge amount of grid data while maintaining the system security at a predefined level. What would be the best platform in which different parties can easily trade energy with each other? How can the grid be immunized against cyberattack and personal information leakage? Promoting trust between participants is another issue coming with the advent of the smart grids. Management of small participants, such as small customers or prosumers, should be managed through a third-party entity like an aggregator. However, the aggregator itself can incur costs for small customers. Furthermore, they may not be reluctant to share their personal information with the other entity.

Blockchain technology could be an efficient solution to address the aforementioned smart grids' problems. The blockchain-based platform consists mostly of a distributed ledger, a decentralized consensus mechanism, and cryptographic security measures. All parties would be able to directly share information and hold the copies of transaction records via their blockchain accounts. Transactions should be performed and confirmed through a set of rules named

Blockchain-based Smart Grids. https://doi.org/10.1016/B978-0-12-817862-1.00001-4

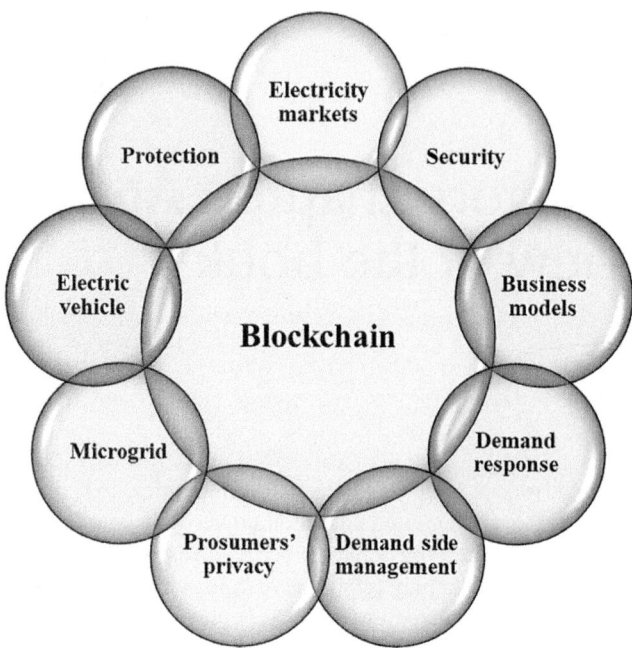

FIG. 1 Applications of blockchain technology in smart grids.

smart contracts. The privacy of information will be highly preserved using hash functions in the blockchain system. Following the advantages of the blockchain-based platform, the main aim of this book is to introduce various applications of blockchain technology in smart grids as presented in Fig. 1. The book chapters are introduced briefly as follows.

Chapter 2, a panorama of applications of blockchain technology to energy, presents the potential applications of blockchain technology in five energy subsectors. The first subsector is energy trading that would be facilitated utilizing blockchain technology. Payments for various sectors can be fully automated through the use of smart contracts, one of the important features of blockchain technology. The second impact of blockchain will be assessed on environmental attribute management including some products such as renewable energy credits, provenance certificates, and guarantees of origin. The use of blockchain technology in energy sectors can incentivize to invest in clean energies as it can solve some challenges related to the current markets. Demand response will be considered as the other application of blockchain technology in energy sectors that will be presented in this chapter. This technology can assist to automate demand response program in a secure way while considering the preferences of customers. The role of blockchain technology in electric mobility will be also discussed in the second chapter. The most effects of this new technology would be on scheduling related to charging and discharging of electric vehicles in the most efficient way. The last energy subsector presented in Chapter 2 is the financing sector. It states that blockchain technology could improve the liquidity of the capital of various projects through the tokenization of assets. Finally, Chapter 2 concludes by introducing 150 companies employing blockchain technology in the field of energy.

Chapter 3, entitled transition toward electricity trading markets blockchain-based, discusses the application of blockchain technology, specifically in electricity markets. It investigates events of the past 10 years, leading to the fourth industrial revolution and the utilization of the decentralized trading platform to incentivize small customers. Blockchain technology will be presented in the chapter so as to satisfy the prosumers' desire for citizen-run democratic energy systems. Afterward the development of the applications of blockchain is discussed. The third chapter continues with future research and milestones with the aforementioned newly emerged technology.

Chapter 4, decentralized operation of interdependent power and energy networks, blockchain and security, describes the transition of conventional networks toward smart grids known as efficient, reliable, secure, and sustainable power grids. Then, blockchain will be discussed to expedite the transition toward decentralization by eliminating the role of an intermediary. The previous literature related to the utilization of blockchain in smart grids is summarized in the fourth chapter. Finally, different frameworks are explained to implement decentralization taking into account the security of participants.

Chapter 5 deals with the application of blockchain technology in different business models related to energy markets. In addition, the roles of various parties in each business model will be analyzed. The business models consist of peer-to-peer, flexibility, over the counter, and crowdsale trading platform. The chapter first represents different parties who are able to take part in the market. Then the interaction between them is assessed. Further, it explains how the blockchain trading platform can speed up the trading while building a reliable platform for different kinds of markets.

Chapter 6, blockchain and its application fields in both power economy and demand-side management, starts with the potential application of blockchain technology in power economy. Besides, it states that demand-side management can also deploy blockchain as a platform to control the consumers' consumption. The architecture, opportunity, and drawbacks of blockchain are fully assessed in the sixth chapter taking into account different viewpoints.

Chapter 7, blockchain-based demand response using prosumers scheduling, represents a two-stage model employing blockchain so as to design a price-based demand response program. The proposed model is considered as a decentralized security-based model in which prosumers do not need to share their private information with any aggregator acting as a broker. In addition, the proposed model is mainly from the viewpoint of prosumers, that is, consumers who are also equipped with renewables to produce energy and manage their consumption. The proposed steps are implemented through a blockchain-based platform to automate the market mechanism. This mechanism aims to motivate prosumers to react to market prices while preserving the prosumers' privacy.

In Chapter 8, energy flexibility is proposed to trade in a secure, reliable, and transparent way in a microgrid environment by employing blockchain technology. In the proposed technique the power presumption data are captured utilizing a blockchain-based distributed ledger technology. The data are gathered from the smart meters, and the smart contracts will recognize programmatically the anticipated energy flexibility of each prosumer, the associated benefits or penalties, and the regulations needed for balancing energy demand regarding grid-level energy production. The blockchain-based consensus mechanism is used to settle the related market and validate the demand response program. The results extracted from several structures of literature show that the blockchain-based distributed demand-side management can match demand and supply efficiently, while the demand response indicator will be tracked with high precision.

The blockchain-based coordination of electric vehicle charging stations is explained in Chapter 9, blockchain for decentralized optimization of energy sources in electric vehicle charging coordination via blockchain-based charging power quota trading. The chapter proposes a two-stage scheme in which the charging quotas are initially allocated to the charging stations. Then, charging stations would be able to trade with each other taking into consideration the demand elasticity of the stations. All the trading and settlements are proposed to be performed in a blockchain-based platform so as to promote trust and benefit from the transparency of this new technology. The effectiveness of the method will be proved in the simulation section of this chapter.

Chapter 10, islanded microgrid management based on blockchain communication, explains the management of a microgrid that will be enabled by the blockchain-based platform. A multiagent market framework is proposed in which the payments between entities will be carried out automatically through blockchain technology. The microgrid is taken to trade with the upstream grid, providing that the generation and demand are not exactly equal.

Another application of blockchain technology in power systems is introduced in Chapter 11, which is entitled blockchain-based protection of DC microgrid. It develops a technique for detecting the fault in DC microgrids. Besides, it tries to isolate the fault, leading to avoiding propagation of damage to the rest of the system. The main role of blockchain technology in this chapter is to encrypt the values sent to the differential relay so as to immunize the system against cyberattacks and communication failures. Finally the proposed method will be tested for a hypothetical DC microgrid.

Chapter 2

A panorama of applications of blockchain technology to energy

Matthew Gough[a,b], Rui Castro[c], Sérgio F. Santos[b], Miadreza Shafie-khah[d] and João P.S. Catalão[e]
[a]*Superior Technical Institute, Lisbon, Portugal,* [b]*INESC TEC, Porto, Portugal,* [c]*INESC-ID/IST, University of Lisbon, Lisbon, Portugal,* [d]*School of Technology and Innovations, University of Vaasa, Vaasa, Finland,* [e]*Faculty of Engineering of the University of Porto and INESC TEC, Porto, Portugal*

1. Introduction

The blockchain technology has the potential to have a significant impact in many sectors in the modern economy, including the energy sector. Numerous use cases for blockchain in the energy sector have been proposed. These include wholesale and retail energy trading, environmental attribute management, aiding demand response programs, and enabling new sources of financing for energy projects.

The blockchain technology has been likened to the Internet in its potential to revolutionize the economy. Other sources say that it is nothing more than a passing fad. This chapter provides evidence on both counts. The potential impact of the blockchain technology is presented, and it is shown that there are cases where the blockchain technology could have significant impacts on the energy sector. That being said, there are a number of challenges that the technology also needs to overcome before it is adopted in a widespread manner.

This chapter is composed of the following sections: an introduction where the context surrounding both the blockchain and the current energy sector is discussed. Following this the blockchain in introduced including the technological background. Then a section of current and past blockchain projects is presented along with a table detailing nearly 150 companies who are active in the energy and blockchain ecosystem. A section of the various limitation of blockchain is then presented followed by a section detailing regulatory aspects relating to blockchain in the energy sector. Following this the major contribution of this work is presented as the various applications of blockchain in energy are detailed.

Blockchain-based Smart Grids. https://doi.org/10.1016/B978-0-12-817862-1.00002-6
5

1.1   Context

The context surrounding blockchain is extremely important in discussing where the technology comes from and where it may be heading in the future. A group of underlying technologies has coalesced to form the blockchain technology, and there are a wide variety of factors that brought these technologies together. This section will discuss the context surrounding blockchain and give the reader a better idea of where the technology originates from and why it has been touted as being a game-changer in the energy sector.

Even before blockchain becomes a hot button issue, the energy sector was beginning to undergo a significant transition. This is shown from a survey conducted in 2013 where 94% of the senior power and utility executives expected either a complete transformation or important changes in the power utility business model by the year 2030 [1]. The impact of the blockchain technology on the energy sector could be very significant as approximately 20% of respondents to a survey conducted by the German Energy Agency believe that the technology will be a "game-changer" for the sector [2]. The context surrounding the energy system is one that can be characterized by the three D's of decarbonization, decentralization, and digitalization, and especially in Europe, there is an overarching goal to empower consumers and put them at the heart of the future energy system [3]. Blockchain can assist in both the decentralization and digitization aspects of the ongoing energy transition.

The existing electricity system is characterized by consumers with little to no control over their electricity use and a paradigm of load-following power plants and very little information available concerning the operating conditions of the system [4]. There is a confluence of factors that are changing how the existing electricity system operates. Chief among these factors is the rapid growth of distributed energy resources (DERs), and this leads to increased customer choice and participation in the electricity market. This paradigm shift has been noted by existing firms, and significant amounts of capital are being spent to upgrade the existing electric system. Estimates suggest that $47 billion was spent in 2016 in upgrades to modernize the electricity system [4]. Despite this significant influx of capital, the current power system still largely functions the way it did in the 20th century. This could be down to the fact that major electricity utilities are normally very risk averse and have to deal with regulatory oversight that could stymie progress [4].

Blockchain helps by reducing two types of transaction costs. The first is the verification of the qualities of a transaction, and the second is the costs of operating a marketplace, which blockchain lowers by removing the need for a trusted third party [5]. By lowering these two costs, blockchain can help create more fluid marketplaces that allow for increased competition, reduced barriers to entry, and lower risks associated with privacy and censorship concerns [5]. Blockchain allows for the transfer of value between two untrusting peers without the requirement of having a trusted third party to verify the transaction [5].

Trust in institutions and on the Internet has been eroded, and this has led to many people questioning the notion of centralized systems with trusted partners [6]. In the past a trusted third party was needed to verify the transaction between two parties. This verification process increased the cost and time taken to complete the transaction, but it was an acceptable trade-off as the third part helped mitigate the risks associated with the information asymmetry and moral hazard problems [5]. The core innovation behind the blockchain technology is the fact that it stores multiple copies of the transaction history and that the copies of the transaction history are connected through a validation mechanism that is secure [7]. Catalini and Gans [5] suggest that the blockchain has allowed for costless verification of transactions. But it may be more correct to say that the requirement of being trustworthy has shifted from the intermediary to the builders of the blockchain code and the chosen consensus mechanism. Therefore blockchain doesn't remove the need for trust; it allows the creation of more trustful relationships [8].

The blockchain technology is expected to have a profound impact on the way that individuals transact among themselves with an oft-cited comparison and has the transformative effect on Internet communication brought about by the introduction of the TCP/IP protocols [5]. Blockchain may help unleash the so-called Internet of Things (IoT). As the number of IoT devices grows worldwide, they will require a secure protocol on which to communicate and transact with each other, and the blockchain may be able to play this pivotal role [5]. Added to this is the fact that the blockchain can transfer details concerning property rights and that these decentralized networks are more resistant to one agent having significant market power. Lowering the barriers to entry in these networks will allow new innovative solutions to be developed, and this will reduce the market power of the existing incumbents. Work carried out by the Research Institute of the Finnish Economy shows that the blockchain technology could play a key role as the interoperability layer in a world with a large penetration of IoT devices [9]. There has been an explosion of interest in blockchain in the past few years. There are numerous conferences and events that have emerged to deal with the blockchain technology and more specifically to investigate the blockchain's potential impact on the energy sector. There has been 24 blockchain-themed conferences in the last 6 months of 2018 [10]. The underlying technology behind the blockchain (cryptography, peer-to-peer networks, and data storage) has all existed for a significant amount of time. What the blockchain does that is so powerful is to combine these three areas of research with various economic incentives, and this allows for the growth of decentralized markets [11].

The use of decentralized systems may also affect the role of the individual in society. For example, in the energy sector, individuals have often had to be passive consumers in which they bought from a centralized source, such as an electric utility. In decentralized and peer-to-peer-based societies, there is scope for the consumer to take on a much more proactive role, and they could become producers of a product as well [12]. The concept of a decentralized electricity system has been spoken about for a considerable time, and the rise of energy

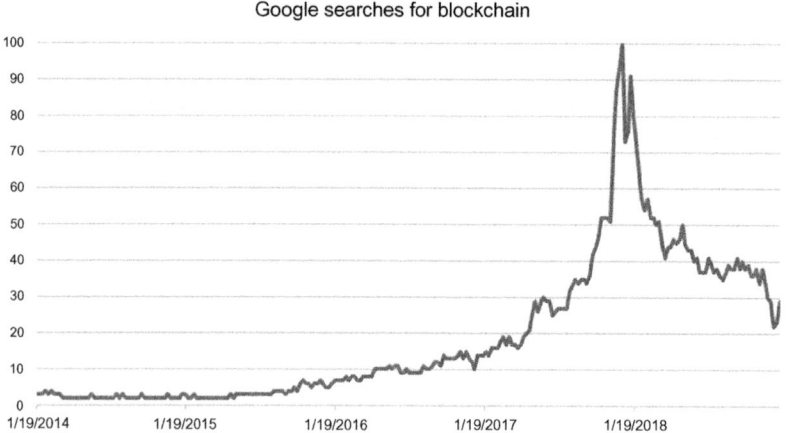

FIG. 1 The considerable rise and decrease in the number of searches for "blockchain." *(Authors' own using data from Google Trends.)*

storage devices and electric mobility, as well as new control systems for demand response, has made the concept of a decentralized grid more tractable.

While there is significant hype surrounding blockchain at the moment, it is thought that blockchain will be truly successful when people use it seamlessly as part of their everyday lives without realizing that they are using blockchain much like the general public uses the Internet today without having to think about the various technologies and protocols that underpin the Internet.

The blockchain technology has received significant attention in the past few years. This can be shown through the Google Trends depicted in Fig. 1. This image shows the popularity of the search topic "blockchain" over the past 5 years. It is a proportional figure where the score of 100 means that this was the peak value of searches for the term and a score of 50 represents a total number of daily searches equal to half of the maximum recorded.

Again, parallels with the growth of the Internet can be made regarding how to ensure future growth of the blockchain technology. The Internet experienced enormous growth when it was user-friendly enough and understandable by the general population. Also, it was predicted that the Internet will usher in a new era of decentralization, but instead, it has become very centralized, and it could be argued that it is undemocratic [13].

In the context of the modern economy with its drive toward the knowledge economy that focuses on knowledge-led growth rather than the traditional means of production, information is playing a key role. Information has become a crucial resource in the modern world [8]. Whoever controls the information can wield significant market power, and the blockchain may be able to democratize information and reduce the market power of existing incumbents that have based their business on the centralized storage and use of information.

1.2   Definitions

1.2.1   Blockchain

There exist numerous definitions for the blockchain, but the general definition of the blockchain is given by Bashir [14]:

> *Blockchain at its core is a peer-to-peer distributed ledger that is cryptographically secure, append-only, immutable (extremely hard to change), and updateable only via consensus or agreement among peers. From a business point of view, a blockchain can be defined as a platform whereby peers can exchange values using transactions without the need for a centrally trusted arbitrator. A block is simply a selection of transactions bundled together in order to organize them logically. It is made up of transactions and its size is variable depending on the type and design of the blockchain in use. A reference to a previous block is also included in the block unless it's a genesis block. A genesis block is the first block in the blockchain that was hardcoded at the time the blockchain was started.*

Another definition is given by Andoni et al. [15]:

> *Blockchains are shared and distributed data structures or ledgers that can securely store digital transactions without using a central point of authority. The data structure is, in other words, a ledger that may contain digital transactions, data records and executables. Instead of managing the ledger by a single trusted center, each individual network member holds a copy of the records' chain and reach an agreement on the valid state of the ledger with consensus. The exact methodology of how consensus is reached is an ongoing area of research and might differ to suit a wide range of application domains. New transactions are linked to previous transactions by cryptography which makes blockchain networks resilient and secure. Every network user can check for themselves if transactions are valid, which provides transparency and trustable, tamper-proof records.*

These two definitions touch on a number of concepts that may be unfamiliar. These concepts are distributed ledger, append-only, immutable, and consensus. These concepts are defined as follows: A distributed ledger is a ledger that is distributed among its participants and spread across multiple sites or organizations and does not reside with a central authority. Append-only means information can only be added, not removed in contrast to read-write databases where information may be removed. Immutable refers to the property of being tamper proof, and while a blockchain can be called immutable, it is technically possible to change the records stored within a blockchain system, but it is extremely computationally taxing, and thus for practical purposes the blockchain system is classified as immutable [14]. Consensus is the method that the various nodes in the blockchain system reach agreement on the various blocks of transactions before being incorporated into the existing blockchain [7]. There are numerous consensus mechanisms that can be used, and these will be described in the coming sections.

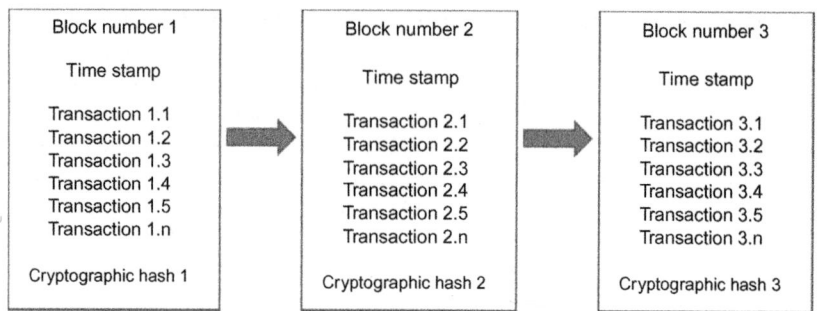

FIG. 2 Typical blockchain system. *(Author modified from J.A.F. Castellanos, D. Coll-Mayor, J.A. Notholt, Cryptocurrency as guarantees of origin: simulating a green certificate market with the Ethereum Blockchain, in: 2017 IEEE International Conference on Smart Energy Grid Engineering (SEGE), in: Presented at the 2017 IEEE International Conference on Smart Energy Grid Engineering (SEGE), 2017, pp. 367–372. https://doi.org/10.1109/SEGE.2017.8052827.)*

Fig. 2 shows the general layout of a blockchain structure after Castellanos et al. [16]. This figure shows a number of blocks each containing a number of transactions. Each block has a sequential number and a time stamp to allow for easy auditing of the blockchain history. Each block also has a unique cryptographic hash that will change should any validated transaction be modified. As each block also makes use of the previous block's cryptographic hash, should any transaction change in any of the preceding blocks, it will change the entire blockchain, and therefore the modification can be easily identified and rectified.

Once a transaction has been requested by a node, it is then broadcast to the rest of the network, which then validates the transaction using the network's chosen consensus mechanism. Once the validation stage is completed, the transaction is then added to a block, and then the block containing a number of validated transactions is added to the existing blockchain.

These various characteristics of blockchain allow digital information to have a value assigned to them. As Joe Ito, the director of MIT's Media Lab, said, "The blockchain makes information look like a thing. It creates the scarcity that you couldn't do on the internet" [17]. This set of technologies that make up blockchain brings together a group of individuals who act in their own self-interest, but together, this group can create an immutable, trustworthy, decentralized system [8].

By prohibiting double spending of the currency, Bitcoin allowed for the creation of a digital asset [8]. In the era before Bitcoin (and the underlying blockchain technology), digital information could easily be replicated and shared (e.g., images and songs). However, blockchain can include an immutable and time-stamped signature of the creator or the original owner, and the spread of that piece of information can now be tracked. It can be said that blockchain (first through the Bitcoin protocol) introduced the concept of digital scarcity.

There are a number of different blockchain types, but the two largest and well known (Bitcoin and Ethereum) will be discussed in the coming sections. They were chosen as Bitcoin is the largest and oldest blockchain system and Ethereum is interesting for the energy sector as it allows applications to be scripted in the blockchain using the Solidity programming language. Ripple is another interesting blockchain application, but as it focuses more on the financial sector, it is excluded from this chapter.

1.2.2   Bitcoin

Bitcoin was introduced nearly 10 years ago in a paper authored by Satoshi Nakamoto (a pseudonym) [18]. Since its introduction, it has become the largest cryptocurrency with an approximate market capitalization of $70 billion as of January 2019. Bitcoin is a collection of peer-to-peer networks, protocols, and software that allow for the creation and usage of bitcoin, a digital currency. Bitcoin can be defined in several ways as it acts as a protocol, a digital currency, and a platform [14]. This section will follow the existing protocol of using Bitcoin, with a capital B when discussing the protocol and bitcoin, with a lowercase b, when discussing the digital currency. Even though Bitcoin is the first and most successful implementation of the blockchain technology, the word "blockchain" never appears in the original paper presented by Satoshi Nakamoto.

Bitcoin's core protocols have so far proven to be very resilient to malicious actors or hacking. Despite being a decentralized system with no central dedicated cybersecurity tools and having a potential $70 billion (as of January 2019) in assets, there have been no successful hacking attempts [8]. However, this does not mean that Bitcoin is inherently safe; there have been instances of hackers gaining access to individual's wallets (where the amount of bitcoin is stored) or obtaining access to the digital currency exchanges where customers can buy and trade cryptocurrency for fiat currency, but this is a software issue relating to the security of the wallet software or the security protocols of the cryptocurrency exchanges and not the inherent Bitcoin protocol. There have also been cases of hackers manipulating the blockchain of smaller cryptocurrencies, such as Bitcoin Gold and Ethereum Classic, but a discussion of that falls outside the scope of this chapter.

Satoshi Nakamoto was not the first to design and implement a digital currency, but what sets Bitcoin apart from the previous attempts is the way that it uses a collection of technologies to make the issue of double spending of the currency extremely difficult. To be clear, double spending is possible on the Bitcoin system, but it is extremely expensive, and this high cost has so far prohibited any individual from successfully double spending a bitcoin. To double spend a bitcoin, an individual would need to control 51% of the computing power of the Bitcoin network to ensure that their version of the Bitcoin blockchain was accepted as the "true" version even if this version of the blockchain contained fraudulent transactions. The costs of such a 51% attack on the Bitcoin blockchain is estimated to cost over $7 billion dollars as of the middle of January 2019 [19].

The manner in which the nodes of a distributed system agree on the correct version of the ledger is known as consensus [14]. Once consensus has been reached, the block of transactions is permanently added to the existing blockchain. Numerous methods exist for reaching consensus, and the exact manner in which consensus is reached can differ depending on the type and nature of the underlying blockchain.

The European Commission recognizes the major contribution of Bitcoin to be the ability to "establish trust between two mutually unknown and unrelated parties to such extent that sensitive and secure transactions can be performed with full confidence over an open environment, such as the Internet" [20].

1.2.3 Ethereum

After Bitcoin the most well-known blockchain is Ethereum. Ethereum also has the second highest market capitalization after Bitcoin with a market capitalization of approximately $16 billion as of January 2019. Ethereum was proposed in late 2013 by then 19-year-old Vitalik Buterin, and he and a group of core developers maintain and upgrade the network.

What sets Ethereum apart from Bitcoin is its built-in programming language that allows it to develop and deploy distributed applications including smart contracts [21]. This allows developers to create applications that run on top of the blockchain and make use of its characteristics. The programming language used within the Ethereum environment is Solidity, and it is a Turing-complete language [8].

To carry out a transaction on the Ethereum blockchain, a user must pay a transaction fee, which is termed "gas." This transaction cost covers the cost of computation required to carry out the instruction, so the more detailed and complicated the instruction, the higher the gas fee [14].

A further evolution of the blockchain system into what is termed Blockchain 3.0 saw the arrival of decentralized applications (DApps) or decentralized autonomous organizations (DAOs). DAOs are defined solely by a collection of smart contracts and allow for operation in a business like environments without the need for human intervention [11]. While this automation of business activities may sound appealing, it could be a very worrisome property if taken to its full capacity. The full automation of the DAO will mean that, once it has been initiated, no one can alter its underlying business logic. Traditional entities adapt over time to better suit their changing environments [21]. A DAO will not be able to do such a thing, and if there is an error in its programming, no one can correct it.

1.2.4 Smart contracts

A key characteristic of the current blockchain ecosystem is the smart contract. While these contracts have emerged as a major defining characteristic of the so-called Blockchain 2.0, the concept of smart contracts has existed for longer than the blockchain concept.

Nick Szabo first defined smart contracts in 1996 as those type of contractual clauses that could be embedded in various aspects of hardware and software to make the breach of the contract inordinately expensive [5]. These contracts can be self-executing and immutable. Another definition of smart contracts is given by Silvestre et al. [22] who define the smart contract as a piece of computer code that verifies certain actions, and should certain criteria be met, corresponding actions are then carried out. Smart contracts can assist in removing the intermediary part in various use cases, and this may lower transaction costs and allow for low-value transactions to take place [15].

The advent of smart contracts raises some interesting questions with regard to the legal profession. The enforceability of smart contracts and even the use of the term "contracts" has raised some arguments. According to Jamison and Tariq [10], a traditional contract requires the following elements to be considered a contract:

1. Offer of the contract
2. Acceptance of the contract
3. Binding agreement to execute a lawful action
4. An exchange of value once the action has been carried out
5. All parties involved should have sufficient legal capacity to enter into the contractual agreement.

Because smart contracts are pieces of immutable, self-enforcing programming code where the authors of the contract can be anonymous or even pseudoanonymous, there are many situations where a smart contract will not meet the criteria listed earlier.

1.2.5   ICO and token sales

The introduction of the various cryptocurrencies has allowed the developers to engage in a new form of raising capital called initial coin offerings (ICOs) [5]. In a typical ICO the developers of the cryptocurrency will allow the sale of a set number of tokens (units of the cryptocurrency) to sellers as a means of investment in the cryptocurrency project or a way to raise capital. There are parallels with the well-known initial public offerings (IPOs), but the ICOs do not generally give the investor a portion of the equity of the project. Rather the investors hope that the cryptocurrency will appreciate in value and thus provide them with an adequate return. ICOs are largely unregulated, and this has led to many fraudulent and poorly supported cryptocurrency projects to launch their own ICO [4]. Many of the ICOs that have already been launched suffer from significant problems including underperformance and failing to advance the project past the conceptual stage [10].

1.3   Characteristics

This section introduces several characteristics of the blockchain technology and assesses its applicability to the energy sector.

There are various types of blockchain available, and the choice of which type to use should be made keeping the characteristics of the organization and problem to be solved in mind.

Blockchains can be split into two types, either public or private. These definitions refer to the one who is allowed to read the content of the blockchain, that is, who can view the content of the transaction history and interact with the blockchain system. Public blockchains are open to anyone to view the transaction data, whereas private blockchains do not allow the general public to view the transaction logs. Blockchains can also be divided along the lines of who can write (add information) to the blockchain otherwise known as validation. Permissionless blockchains allow any node to write and commit information to the blockchain, while permissioned blockchains only allow certain nodes to write to the blockchain [7]. Permissioned ledgers generally have their members known to each other, and this helps facilitate trust in the group of nodes. These types of ledgers do not need to use a distributed consensus mechanism, but they can rather use a predetermined agreement protocol [14]. The difference between a permissioned and permissionless blockchain is shown in Fig. 3 [15]. In the left-hand network, each node has the ability to read and write to the blockchain (shown by the images of the computer and the checked paper). In the right-hand side, only the red nodes can validate transactions. Permissionless networks can be thought of being more decentralized and democratic than permissioned networks, but each network has its advantages and disadvantages, and the choice of network architecture should be made after careful consideration of the project's specifications.

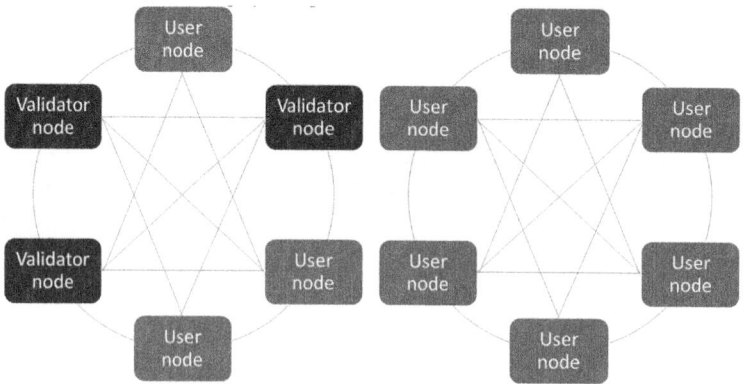

FIG. 3   Difference between permissionless *(left)* and permissioned *(right)* blockchain systems. *(Author modified from M. Andoni, V. Robu, D. Flynn, S. Abram, D. Geach, D. Jenkins, P. McCallum, A. Peacock, Blockchain technology in the energy sector: a systematic review of challenges and opportunities, Renew. Sustain. Energy Rev. 100 (2019) 143–174. https://doi.org/10.1016/j.rser.2018.10.014.)*

There are examples of hybrid blockchains that have characteristics of both public and private chains. These blockchains are called consortium blockchains and are often targeted at enterprise use. They include Hyperledger, Corda, and Quorum.

A large factor in the performance characteristics of a blockchain (scalability, transaction speed and finality, security, and use of resources) is the method used by the blockchain to reach consensus [15]. Choosing the correct consensus mechanism is critical as the mechanism has to be flexible and responsive enough to allow the blockchain to work optimally, but it is the consensus mechanism that also has to protect the blockchain from faulty or malicious nodes, and thus the consensus mechanism has to be resilient enough to protect the blockchain system.

To an end user a distributed system seems like one unified system, but in reality the different nodes in the system have to coordinate their activities so as to provide a common outcome. This coordination between the various nodes and the systems' fault tolerance are the main challenges for distributed systems. Fault tolerance refers to the ability of the network to sustain some nodes becoming faulty and still remain available.

Distributed systems are governed by the CAP theorem, also known as Brewer's theorem. This theorem posits that a distributed system cannot have consistency, availability, and partition tolerance simultaneously [14]. Consistency requires that each node in the system have the latest version of the data. Availability requires the system to be operational and functional. The partition tolerance requirement states that the system is capable of surviving a failure of a group of nodes within the system. A common method used to achieve fault tolerance is replication, and consensus mechanisms are used to ensure that the consistency aspect of the CAP theorem is satisfied.

Blockchains choose to concentrate on availability and partition tolerance, and consistency is not achieved simultaneously with availability and partition tolerance, but it rather is achieved over time, and this is termed eventual consistency [14].

As blockchain has evolved over the years, there can be three broad categories into which blockchain projects can be classified into according to their complexity and level of autonomy [7]. Cryptocurrencies (such as bitcoin) can be classified as being part of the Blockchain 1.0 family. The introduction of smart contracts is the key defining characteristic of Blockchain 2.0 applications where the first set of autonomous actions can be carried out by the blockchain should certain criteria be met. Blockchain 3.0 is characterized by decentralized autonomous organizations, which take the concept of smart contracts to the next level. These are organizations that are solely run on a collection of smart contracts and a high degree of autonomy. Currently, Blockchain 1.0 and 2.0 applications are most common, and Blockchain 3.0 applications will require further maturation of the technology [7].

The design and operation of electricity networks are classified as a natural monopoly where the high capital costs, economies of scale, and other barriers to entry provide an advantage to the first mover in the system and restrict access to the market. Often, these first movers are then regulated so as to ensure that consequences of this type of market failure do not impact the customers. This often requires that these agents have unique responsibilities for the correct operation of the networks. These requirements are at odds with the decentralized nature of blockchain systems, and finding a solution to this issue is a key aspect if the blockchain technology is to be successful in the energy sector.

The hype around blockchain can be thought of as both a help and a hindrance. Blockchain can help move the loci of trust from a central platform operator to trust in the underlying blockchain protocols and programming code [5]. Because blockchain reduces networking costs, it affects the issues of market power of the intermediary, privacy risks, and risks associated with censorship. This may reshape the architecture of the electricity market and create new opportunities for new entrants to take advantage of the new architecture [5]. Blockchain offers the following key characteristics that will be beneficial to recording transactions. The general use case of blockchain has the following characteristics:

- A database is needed to order and record transactions
- Multiple users need to use the database to add transactions
- The ordering of the transactions is crucial
- There may be malicious actors within the network, which limits trust in the network
- There is a need for disintermediation [12]

If there is no need for disintermediation or there is a need for a trusted third party, the use of a blockchain-based system may not be needed as there are significant computation and communication costs associated with implementing a blockchain system [23]. Blockchain is thought to be most useful in a situation where the transaction does not involve a physical exchange [7].

As has been described in the earlier sections, the blockchain protocol has shown to be very resistant to hackers and other malicious actors. The weak link, therefore, becomes the device connected to the blockchain, and a further weak link is the user of the device.

There are some characteristics of the blockchain ecosystem that may allow for the creation of systems that steer communities toward socially beneficial actions, and this might help solve issues relating to the tragedy of the commons problem [8]. This aspect of blockchain has received significant attention recently, and this has coined the term "cryptoeconomics" [24]. In short, cryptoeconomics combine cryptography and economic principles to create decentralized peer-to-peer networks that are robust and can thrive in an environment where there are malicious or adversarial peers [24].

2.   Current and past projects

The blockchain ecosystem has grown at a rapid rate in the past few years. This section will examine some of the most noteworthy projects that have emerged. Luke et al. [7] states that as of March 2018 the number of organizations working on blockchain in energy numbered 122 and there were 40 projects announced. This number can only be expected to grow. These are listed in Appendix 1 at the end of this chapter.

One of the most cited projects of blockchain in energy is the Brooklyn Microgrid designed by LO3 Energy. This project consists of a microgrid energy market situated in Brooklyn, New York, and gained significant attention by performing the first-ever blockchain-based peer-to-peer electricity transaction [25]. The project comprises less than 60 participants, and the participants are connected to the existing distribution grid, and when the participants transact, they are not transacting electricity but rather a form of renewable energy certificates [4].

Another existing project is Grid+, which is based in Texas and aims to give residential consumers better access to participate in wholesale electricity markets [4]. It still has a considerable way to go before reaching its final goal of helping residential consumers manage their bills, trade electricity, and offer their distributed energy assets to help manage the distribution grid; Grid+ can currently assist consumers to manage the component of the bill related to the wholesale costs of electricity, and this is often a small component of the overall bill [4].

A startup based in the United Kingdom, Electron, is aiming to create a flexibility market for electric power using the blockchain. NRGCoin, an initiative, aimed at helping support the integration of renewable energy by providing incentives for the local production and consumption of renewable energy [23].

An example of a project funded by both industry and government is the Irish EnerPort project. This project seeks to develop peer-to-peer energy trading networks between microgrids [26].

While the startup companies who are aiming to disrupt the energy sector by using the blockchain technology may receive a significant amount of media attention, those companies who seek to work along with incumbent entities are most likely to succeed [4]. The key will be to see the blockchain act more of a platform to enable other information technologies and less of an instant solution [4].

The role of blockchain in energy has not only been explored by startup companies. Incumbent companies in the energy sector have begun to explore how they can use the blockchain technology. A key example of this is the Energy Web Foundation. This foundation aims to accelerate the adoption of the blockchain technology in the energy sector and counts among its affiliate members Centrica, Duke Energy, Engie, E.ON, Shell, TEPCO, and GE [27].

The blockchain system designed by the Energy Web Foundation has a confirmation time of between 3 and 4 s and can handle several thousand transactions per second [15].

The Enterprise Ethereum Alliance (EEA) was launched in March 2017. This initiative brought together various blockchain startups, research centers, and different Fortune 500 companies. The EEA sought to create an open-source standard for the blockchain [21].

Another example of a large-scale collaboration is the Hyperledger project that was formed in December 2015 and aims to create an open-source distributed ledger framework to help boost interoperability between blockchain applications and systems [14]. The Hyperledger project is composed of a number of frameworks and tools. There are five frameworks and five projects as of January 2019 [28].

There are a number of large organizations offering Blockchain as a Service (BaaS). These include Microsoft's Azure, IBM's Bluemix, SAP's HANA and Leonardo blockchain projects, and recently Amazon Quantum Ledger Database and Amazon Managed Blockchain. These tools help ease the difficulty of interacting with a blockchain system, and these projects will be crucial in increasing the mass market adaptation of the blockchain technology.

It is not only large corporations that are investigating blockchain, but also some political and governmental organizations are investigating the possible use cases of the technology. The largest is the European Union's Blockchain Observatory and Forum. The goal of this forum is to examine the potential impact of the blockchain technology and potential uses of the blockchain technology for the citizens of the European Union [29]. A number of European countries are attempting to position themselves on cutting edge of blockchain technology. These include France, Germany, Austria, Lithuania, Estonia, Spain, and the Netherlands [29]. Funding totaling €30 million has been dispersed by the European Union with an additional €300 million in possible future funding. A report showed that 60% of large corporations are exploring blockchain technology and nearly 90% of international banking executives have initiated exploring the possible impacts of blockchains or distributed ledgers in payment applications [29].

During the course of writing this chapter, a detailed list of various companies and initiatives working with blockchain in energy was developed. This list is presented in Appendix 1 and lists the primary field of each company and the location of nearly 150 companies and initiatives.

3.   Limitations

While there have been many examples of the potential benefits of incorporating the blockchain technology in the energy sector, there are still some limitations that will need to be overcome before the technology can make a significant impact in this sector.

While not a limitation of the blockchain technology itself, there is an inherent limitation in applying the technology to the electric sector. The electric sector is often characterized by large-scale, centralized systems that make use of both economies of scale and economies of scope. These factors combined with the risk-averse nature of many electric utilities could limit the speed and scale of the blockchain impact in the sector. The fact that there is often a physical transaction coupled with a financial transaction in the energy sector raises some issues with regard to blockchain's potential impact on the sector. This is especially true in the electric power sector as, once injected into the grid, it is very difficult to control and track the electricity from the supplier to the consumer.

Public and permissionless blockchains would allow the highest number of people to join a blockchain system, but the trade of this increased size is the transaction speed and the high costs of proof of work consensus mechanisms that have dominated the public and permissionless blockchain ecosystem. This trade-off has been termed the "scalability trilemma," and it states that a blockchain can only have a maximum of two of the following three characteristics: decentralization, scalability, and security [7]. Innovation and research will be needed to overcome these challenges. There is some progress being made with different consensus mechanisms such as the Tobalaba network created by the Energy Web Foundation that uses a proof-of-authority consensus mechanism.

There is a school of thought that states that the security of a blockchain can only be tested once it has grown to such a size that the reward for hacking the system becomes attractive [7]. This could be a challenge for an early project using blockchain in the energy sector. The strength of the network can only be fully tested once it has grown to such a size that it becomes the target of a coordinated attack. A successful attack on such a system would have a significant impact on society given the critical role electricity plays in the modern economy.

The allocation of legal and technical responsibility of the blockchain also may become an issue should unforeseen events take place such as a security breach [7]. This issue is often compounded due to the lack of a hierarchical structure of the developers of the blockchain. There is often heated debate among the programmers when certain modifications of the underlying code need to be carried out. This problem could also highlight the lack of flexibility of a blockchain once they are deployed, and this issue is made more difficult with the development of DAOs.

The issue of using a blockchain as a register of physical assets could be an issue should a fork or split of the blockchain take place [7]. Different people having ownership rights of the same asset at the same time could lead to difficult legal issues.

In cases where smaller systems composed of individuals who already have a degree of trust in each other, the benefits of running a decentralized system of distributed ledgers do not always perform better than a centralized database [11].

Maintaining trust in the blockchain system is key for its continued survival, but the volatility in the prices of various cryptocurrencies shows that trust and confidence in these systems fluctuate significantly [7]. The volatility of the cryptocurrency prices is also another factor limiting the adoption of cryptocurrencies as a widespread medium of exchange. Splits and forks in the blockchain could also introduce further uncertainty and decrease the trust in the network, and this necessitates very good governance and change management strategies to reduce the possibility of a fork [5].

Should the energy system develop into one characterized by millions of decentralized energy resources making multiple transactions per day, the need to store all of these transactions in the blockchain may pose a technical challenge because it would be very difficult for each node of the network to store a complete copy of all the transactions [30].

A partial remedy for the scalability issue could be using off-chain transactions [12]. There are also initiatives such as the Raiden and Lightning networks to improve the speed of the Ethereum and Bitcoin networks, respectively. Sharding or partitioning the blockchain into a number of smaller chains is also another proposal to make the blockchain more efficient [12], although sharding also has its own challenges that need to be overcome before it can be implemented successfully.

As of January 2019 the Bitcoin blockchain is processing between three and four transactions per second on average, and the Ethereum is processing approximately seven transactions per second [31]. This is much smaller than comparable transaction processing networks such as Visa or Mastercard, which handle roughly 5000 transactions per second.

As the blockchain ecosystem is still in a nascent stage, the choice of consensus mechanisms and other aspects relating to the blockchain architecture could be difficult for developers to make. So far, there has not been a dominant consensus mechanism or system architecture to emerge, and this may hinder the progress of blockchain developers as they may not know the pros and cons of each system choice [15].

State channels may offer a way to increase the speed of a blockchain network. These involve opening up a dedicated side chain to record multiple transactions between two parties, and once the trading has been completed, the final accounting is then added to the main blockchain. In this way, there is only one final transaction added to the main chain rather than several intermediate transactions between the same two parties [14].

One blockchain that aims to tackle the issue of throughput or transactions per second is the EOS blockchain. The developers of this blockchain aim to reach between 6000 and 8000 transactions per second [32]. As of January 2019 the EOS network has reached a maximum of 3996 transactions per second [33].

The complete removal of trusted intermediaries may not be desirable as they often play other roles in society. Therefore the use of the blockchain may force a reinvention of these trusted third parties to concentrate on their other roles in society [7].

There have been counterintuitive cases where the success of a blockchain has been a hindrance to it as well with the new-found popularity becoming a major issue to the underlying network. An example of this is the CryptoKittie trend that disrupted the Ethereum network toward the end of 2017. A CryptoKittie is a unique digital pet generated by the application's code when two other CryptoKitties are paired together and "breed." Each CryptoKittie is unique and recorded on the Ethereum blockchain. The trend grew rapidly and put severe pressure on the Ethereum network, which slowed the entire network [34]. The explosion in popularity of the CryptoKittie did raise the profile of Ethereum and made it seem more perceivable to the general public. But this rise in profile came at a significant cost, and in December 2017 CryptoKitties made up to 20% of the total Ethereum traffic, and this traffic severely slowed the remaining Ethereum network. The CryptoKittie trend raised doubts on the potential of the Ethereum network to handle vast numbers of transactions from distributed applications, and the network will require further development before it can handle a significant number of transactions [34]. The trend of CryptoKitties lead to a sixfold increase in the volume of Ethereum transactions and showed that Ethereum is currently not able to handle the volume of transactions that large-scale applications will bring [35]. The issues caused by CryptoKitties also had some positive outcomes as more developers realized that there would need to be a significant increase in the transaction processing capacity of the network.

There also needs to be significant work on developing standards and other means of ensuring interoperability between two or more blockchains [14]. There has been work carried out in this field, most notably the development of the ISO/TC 307 technical committee tasked with the standardization of blockchain and distributed ledger technologies [14]. The Hyperledger project combined with other open-source efforts such as the R3 project and the open-chain standard are working to ensure a common pool of blockchain tools and protocols.

Self-governance of blockchain systems is also a major challenge facing developers today. The ideals of decentralized, immutable, and pseudoanonymity do not generally make for the easily establishing a robust governance structure and to do this without sacrificing the ideals of a project based on open collaboration between a group of core developers.

4. Rules and regulations

There are a number of rules and regulations that already influence the impact of blockchain, not only in the energy sector but also across other sectors. The pseudoanonymity offered by blockchains may fall foul of existing know-your-customer (KYC) and anti-money laundering (AML) rules. These rules may require organizations to expend time and effort to link blockchain accounts to real-world identities [5]. This may require a centralized authority to monitor and verify the names and addresses of the users of a blockchain system that will go against the spirit of decentralized systems.

Regulators not only do have a responsibility to oversee and safeguard current applications of blockchain but also do have a powerful role in shaping the future growth of blockchain applications. Proactive policy to assist the immature blockchain technology is needed to provide assurances to developers and financers for the blockchain ecosystem to grow and develop. That being said, there is a real need for regulatory policies, which seek to minimize the risks of current users of the blockchain, such as regulating ICOs so as to restrict the number of poorly designed or even fraudulent projects. An example of such policy is the European Union's Digital Single Market and the Declaration on the European Blockchain Partnership [7]. Another example is the joint European Committee for Standardization (CEN) and the European Committee for Electrotechnical Standardization (CENELEC) to develop the CEN-CENELEC Focus Group on Blockchain and Distributed Ledger Technology whose main activity is the development of the ISO/TC 307-Blockchain and Distributed Ledger technologies standard.

Standards will be necessary as new equipment may need to interact with demand response programs and these types of equipment (household appliances and commercial and industrial equipment) generally have lifetimes of a few years, so they will need to be prepared for future developments in decentralized energy markets.

There needs to be a balance between ensuring that households and communities have an adequate supply of energy, including in extreme situations, and also promoting other regulatory goals such as promoting the use of low-carbon energy sources. There have been several examples of regulations being passed to allow self-consumption of energy (in France) and allowing the limited peer-to-peer energy trading (in Germany, the Netherlands, and the United States) [12]. Regulatory policies also shape the business models of firms operating in the electric power sector [1]. In some cases, especially distributed energy resources, business models are driven more so by regulatory and policy factors than by technological considerations [1]. There are also issues of creating regulatory dependence among business models, and thus regulations need to be carefully considered.

Alongside regulations, it is very important to get the tariff design right as this could have significant impacts on the sector and society at large. For example, poorly designed tariffs are thought to have forced electricity consumers in the United Kingdom to pay an additional £1.4 billion pounds per year over the period 2012–15 [36].

How the blockchain fits into the European Union's General Data Protection Regulations will be an interesting and crucial dynamic to resolve in the coming years. The GDPR states that in certain case user data should be anonymized or erased. How this can be done in the blockchain with its permanent and immutable record of transactions may be challenging [4]. The GDPR also allows individuals to have access and control of their data in certain cases that again run against the principles of the blockchain technology [37]. Protection of user data lies at the heart of both the blockchain technology and the GDPR, but the two attempt to solve the problem in drastically different ways, and as such, there have been attempts to interpret the GDPR in a flexible manner so as to allow the blockchain to operate in the European environment [37]. The GDPR also states that data can only be sent to third parties who are outside of the EU if that jurisdiction has equivalent standards for data privacy as the GDPR [7]. This may be an issue for public blockchains with nodes spread across numerous locations as the blockchain cannot direct data to certain locations.

Blockchain technology may be faced with some short-term reputational impacts due to the perceived issues of bitcoin (and other cryptocurrencies) facilitating illegal activities. This may be difficult to overcome until the public learns more about the blockchain technology and are exposed to successful projects that actively shun the shadow economy that has been known to use cryptocurrencies for illegal activities.

From a policymaking perspective, it will be key for policymakers to understand the blockchain technology before attempting to make policy to regulate and control it. Following this, it will be imperative that efforts be made to standardize the technology and increase the interoperability of the various networks. Regulatory sandboxes or pilot projects with relaxed regulatory constraints should then be implemented to see how the technology works in the real-world (although small-scale and controlled) environment [4].

Regulators should be adaptable and take lessons from other sectors that experience the rapid proliferation of digital technologies that challenged the status quo of that sector (such as the telecommunications industry) [10].

ICOs have also come under the increased scrutiny of regulatory bodies, especially the Securities and Exchange Commission (SEC) in the United States. The issue of whether or not an ICO is considered a security was a major question surrounding the ICO environment. In March 2018 the head of the SEC confirmed that ICOs were considered securities and thus had to follow the usual rules and regulations of the asset class [38]. Cryptocurrencies do not meet the security definition as stated by the head of the SEC [38].

5. Applications of BC to energy

This section briefly touches on the various subsectors within the energy sector that may be impacted by the blockchain technology. This section only provides a brief overview of the various applications of blockchain technology as the remaining chapters of this book examines each of the sectors in more detail.

When assessing the potential impact of the blockchain technology on the energy sector, there are five areas within the energy sector that may be affected

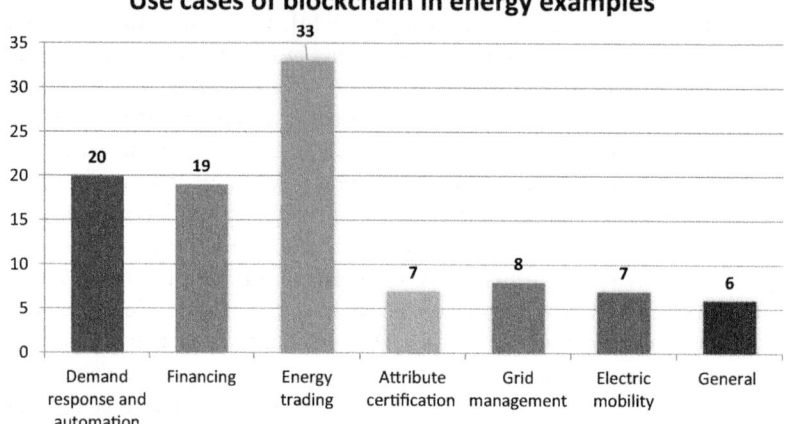

FIG. 4 Breakdown of the different use cases of examples of blockchain in the energy sector. *(Author modified from M. Andoni, V. Robu, D. Flynn, S. Abram, D. Geach, D. Jenkins, P. McCallum, A. Peacock, Blockchain technology in the energy sector: a systematic review of challenges and opportunities, Renew. Sustain. Energy Rev. 100 (2019) 143–174. https://doi.org/10.1016/j.rser.2018.10.014.)*

as shown in Fig. 4. These are energy trading (wholesale and retail markets), attribute management, demand response, electric mobility, and financing of projects within the energy sector. This section will examine each of these use cases in detail and then provide a short section detailing general considerations regarding the impact of the blockchain technology in the energy sector.

5.1 Energy trading

One of the most widely cited applications of blockchain in the energy sector is its potential impact in the energy trading sector. Blockchain could speed up the payment for various services and allow for payment to follow automatically as soon as the transaction has been completed. This could be achieved by using smart contracts. Traditionally the payment process has lagged behind delivery quite significantly, and a whole business sector has emerged to deal with the payment and settlement process.

The rapid settlement of transactions will become increasingly important because the volume of transactions is expected to grow as more individual households are in a position to produce, consume, and trade electricity [39]. Currently the trading process involves numerous transactions between numerous actors, and this means that the process is ripe for the possible disintermediating effects of blockchain technology.

Energy trading is a fairly new phenomenon as there was very limited trading taking place before the year 2000. Historical demand figures were used to forecast the demand for electricity with the generators carrying out voltage and frequency control [40]. The liberalization of the European energy markets created a transition from the vertically integrated utility toward a more horizontal market with a large number of both suppliers and buyers of energy products, and this led to a significant rise in the number of transactions carried out. Along with the increased number of transactions, there was a rise in the number of products being traded including both long-term (futures-based) and short-term deals (spot market). In the futures markets, there are both physical and financial products being traded. The financial product sector is where the blockchain could work best as these trades do not involve a physical transfer of goods. Merz [40] suggests that the initial use case for blockchain technology in this sector could be as part of the communications network handling the trades. This could be especially useful in standardizing the information flow between traders. Should the markets make use of one blockchain to coordinate the trades, all traders will need to provide the same information for their trades, which could increase the transparency of these trades. The blockchain could help remove intermediaries in this system, which are often cumbersome and slow the process down.

The liquidity of the market will be improved should the time taken between the delivery of the energy assets and the associated settlement is reduced [40].

Those energy trading projects that state that they allow their consumers to choose where their electricity comes from will never be able to actually allow their customers this choice. Their promise does not take into account the nature of the flow of electricity in a network and Kirchhoff's circuit laws [40]. However, should the energy markets incentivize the consumption of locally produced electricity, an increase in the development of generation assets near to demand centers could be seen as opposed to the previous paradigm of locating the generation assets near to their fuel resources (mostly coal and natural gas).

The current way of handling the trading can result in high transaction costs and operational inefficiencies. A possible example of the blockchain in energy trading has been developed by Ponton. This is the "EnerChain" project, which is a decentralized version of a clearinghouse using the blockchain technology [7].

Some of the longest lags between delivery of energy and the associated payment occur within imbalance settlement markets with payments taking up to 28 months to be settled [15]. Smart contracts could be utilized to automatically carry out the payment once the delivery of the energy has been received and verified. Smart contracts could also create a more transparent market and reduce other inefficiencies within the existing markets. Accordingly the potential impact of blockchain in this sector has been realized by investors with 57% of the capital raised for blockchain in energy projects being allocated to projects looking to increase the speed taken to verify and execute transactions [7].

Another aspect of energy trading in which blockchain may have a significant impact is the recording of data associated with the condition of various assets within the network [7]. Additionally, due to its inherent redundancies, blockchain could help protect against the risk of cyberattacks on the electricity network. This risk is only expected to grow as more devices are connected to the Internet and operate at the edge of the grid where it may be difficult to enforce existing cybersecurity measures [6].

However, using the blockchain in energy trading markets still has some issues to work out. It may still be very difficult to track the flow of electricity in a network so as to verify the transaction between a supplier and consumer of the electricity [41]. Additionally, even if the market allows energy trading agreements to be set up between two parties, these agreements will need to be approved by a central authority once a technical feasibility assessment has been done to prove that the network can handle this transaction. Should the network not be able to handle the agreed transaction, it may be difficult to renegotiate the agreement if smart contracts are used.

Using the blockchain technology to enable peer-to-peer markets can increase the customer's participation in that market and put the customers at the heart of the energy transition. In these markets, individuals could offer to sell their excess generation or market their flexibility to other market partici- pants. This could help in increasing the penetration of small-scale renewable energy sources and increasing the flexibility of the market. Various use cases for blockchain in energy trading have been presented and include peer-to-peer trading in microgrids, bilateral agreements between producers and consumers, demand response actions, coordination of virtual power plants (VPPs), manage- ment of the grid, energy storage management, aiding in the control strategies of DERs, community energy initiatives, and coordinating a power plant portfolio [15]. It is unlikely that the energy system will ever be fully decentralized as there will always be a need for a central authority to manage and control the distribution and transmission grid. Blockchains could also expose consumers to the real price of energy. This could lead to more rational energy choices or increased participation in demand response activities [15].

The roles played by the different actors in the energy system may also change. The owners of the transmission and distribution (normally the TSO and DSO) will still be responsible for operating and maintaining the physical networks, and they will need to be compensated for that plus their roles as sys- tem stabilizers. Each node in a peer-to-peer energy market will need to be responsive to the needs of the grid and react accordingly. These grid needs may include network conditions, prices, and balancing supply and demand [42]. This will increase the onus on the owners of the DERs to provide timely and accurate information to the system operator.

These factors combine to make it very unlikely that a fully decentralized electricity network that makes use of peer-to-peer energy trading will develop in the next decade. This is because the existing grid, while flawed, does provide services to its customers that a decentralized network may struggle to achieve [4]. As the roles of actors within the electric power system change and evolve over time, there may be scope for blockchain systems that work with the incum- bent utilities to help manage the grid [4]. Blockchain is likely to reduce the barriers to entry for individuals to participate in an energy trading market; this increases both transparency and liquidity of the market. Blockchain also allows for more flexible generation portfolios with a diverse set of resources and numerous transactions occurring between the market participants [43].

5.2   Environmental attribute management

Using blockchain systems to help manage the environmental attributes (systems that reliably verify the origin of various units of electricity), this sector encompasses markets for products such as guarantees of origin, provenance certificates, and renewable energy credits). The generic term of Energy Attribute Certificates (EACs) will be used in this section. The EAC sector provides a solid foundation for blockchain systems. The existing systems are generally complicated and require numerous trusted third parties to ensure compliance, open to fraud and errors; the markets are thus often slow and cumbersome with high barriers to entry, and there is no physical transfer of goods in this market. Rather the EAC markets revolve around providing verifiable proof that a particular individual has the ownership rights of a particular attribute. Challenges associated with the current system also include having to rely on costly manual audit processes, the limited geographical scope of existing systems, and opaque management practices [7].

Using blockchain to run these markets may increase the amount of capital invested in clean energy technologies and therefore help reduce emissions relating to the energy sector [4]. The European Union Emissions Trading System (EU ETS) requires nearly 11,000 thermal power plants with capacities of over 20 MW to purchase emission certificates to account for their related CO_2 emissions [44]. Lowering the transaction costs of participating in such attribute markets could also allow for the introduction of smaller power plants and even include other sectors of the economy that have traditionally been left out of such attribute markets such as transportation or residential heating [44]. Costs of interacting with the current markets can run into thousands of euros per year, and this acts as a large barrier to entry for many smaller actors who would otherwise like to participate in the market [16].

There are concerns that the future of attribute markets is not so positive in the long term as, should the global economy switch to 100% renewable energy sources, there will not be a need to record the environmental attributes of each unit of electricity [44]. That being said, attribute markets are an effective tool to enable that transition.

Like in other sectors, blockchain technology holds the promise to reduce transaction costs. In situations where the allocation of externalities plays an important role (such as emission trading schemes), lowering the transaction costs can increase the efficiency of the outcome and bring it closer to being the Pareto efficient outcome as is described by the Coase theorem [45]. Quantifying and monitoring emission target are a crucial pillar in any climate change mitigation policy, and the blockchain can help to quantify and monitor these emissions.

There are examples of initiatives using the blockchain technology in the field of monitoring and verifying environmental attributes of energy. One of these examples is SolarCoin, which is a cryptocurrency that can act as a guarantee of origin system by using proof of generation [16]. Proof of generation means that, for every 1 MWh that a solar PV system generates, the owner of the system is awarded one SolarCoin, and these SolarCoins can be traded at a later stage.

There is also a pilot project being run in Thailand to examine the effectiveness of blockchain applications in this sector. The project is being developed by South Pole, ixo Foundation, and Gold Standard. The project consists of 10 solar farms and seeks to examine the role of blockchain in the monitoring and verification of information relating to carbon credits [46].

5.3   Demand response

Another major part of the energy system that the blockchain could impact is demand response (DR). Demand response programs could be encoded in smart contracts and shared among the participants in a given location. This could help automate the demand response program as each individual could set their preferences and advanced machine learning algorithms could help predict the load profile of each household so as to provide the system operator with the most up-to-date information regarding the available demand that could be utilized. In this example, smart contracts can operate as a decentralized control strategy. There may also be markets for ancillary services (such as voltage regulations and reactive power control) where the blockchain system and smart contracts could help activate, monitor, and report the results in a verified, immutable, and automatic manner. Smart contract with their ability to operate automatically and very quickly may help issues relating business models relating to residential demand response [1]. Coordinated dispatch of demand response from DER devices could help utilities defer expensive upgrades or expansion of the distribution grid, and the demand response effort of DERs could also assist in grid management by providing ancillary services [4].

It is estimated that there could be approximately 21 billion smart devices connected to the Internet by the year 2020 with nearly a large portion of those devices having a direct impact in the energy sector (such as smart meters, thermostats, and other devices) [2]. This number of devices will provide a significant amount of demand response capacity if it is harnessed and utilized correctly. The concept of the smart grid requires that devices communicate between themselves in an organized and automatic manner and these devices could respond to different signals related to energy price, grid needs, or renewable resource availability [15]. The number of devices and transactions would make centralized systems unwieldy and inefficient, and thus decentralized decision-making and control could provide significant benefits [15].

There are a number of issues that need to be overcome before this future can be realized. It will be necessary to develop power electronic devices that can communicate with each other and record information to the blockchain system, and these devices will need to be able to measure the demand of various devices in each household. Combined with this technical challenge, there is a societal challenge that will also need to be addressed, and this is the issue of data privacy. It is unclear how comfortable consumers would be sharing the data related to the usage of their household devices [15].

In a similar vein to the previous two use cases of blockchain in energy, the barriers to entry into demand response markets can be lowered, and this would significantly increase the size of the market. For the United Kingdom, this larger demand response market could unlock an estimated £1 billion in savings, a net savings of £1.4–2.4 billion per year by 2030, and an overall saving of £17–40 billion by 2050, and these cost savings are reached through a combination of reduced capital expenditure on generating assets, lower operations and maintenance costs, reduced network upgrades, and lower costs for security of supply [47].

5.4 Electric mobility

Blockchain technology also has the ability to play a major role in electric mobility (EM) sector [15]. In the EM sector the most discussed roles of the blockchain technology are recording the transactions relating to charging of the vehicles and the creation of smart contracts to help electric vehicles offer grid services to the distribution system operator. The decentralized nature of transport (with numerous vehicles, drivers, and dispersed charging infrastructure) makes applying the blockchain technology an attractive option. Using blockchains to help manage electric mobility services will reduce the need for a central authority to manage the vehicles (if they are owned by a ride-sharing platform) and the charging infrastructure and improve the system resilience, and the blockchain can also help create efficient markets in this sector [15]. The major concerns of using this technology in this sector are related to the privacy and security of data. The electrification of mobility is helping to bring together the electric power and transport sectors [4].

5.5 Financing

Blockchain technology could increase the size and liquidity of capital available for various projects in different sectors, including projects in the energy sector. It can do this by lowering the barriers to entry to these markets and thus making them available to smaller investors who would normally be locked out of such capital pools. Blockchain could also impact the financing of energy projects through the so-called "tokenization" of assets. Tokenization converts the ownership rights to a particular asset to a digital token that can then be split up into a number of different parts so that individual investors can own a portion of the physical asset. This way a single investor can buy a portion of a tokenized renewable energy project, and their return will be proportional to the percentage of the asset that they own. Apart from tokenization of assets, cryptocurrencies can be designed to reward certain types of behavior, and this could be used to spur individuals to act in a beneficial manner when they might otherwise not do so.

5.6 General considerations

The blockchain also has the potential to impact certain activities that cut across numerous areas in the energy sector. These activities include increasing the speed and accuracy of data recording, lowering the price of verification of data, reducing the need for a number of intermediaries, and automation of existing processes. This will mean that, by using the blockchain in the energy sector, costs could be reduced, and this could translate into lower fees and levies for the end customer.

Tracking shipments of oil and gas could be made easier using the blockchain as the technology can help digitize documents and improve supply chain management. This has already been seen in other sectors such as the maritime and aviation. In the overlap between the maritime, aviation, and energy sectors, the use of blockchain technology could provide significant benefits when it comes to dealing with bunker fuel shipments. There are large amounts of high value product in the bunker fuel market, and it is necessary to accurately track and record transactions to reduce the possibility of fraud occurring [48].

The blockchain may help the electricity network become more resilient, secure, and reliable as decentralized systems do not have single points of failure as it is common in large-scale centralized systems [7,39].

The blockchain has been touted as a revolutionary technology that will drastically transform the way we carry out our daily lives, but the technology should never be viewed or used in isolation. It should form part of a diverse set of technologies that should be tailored to suit the characteristics of the problem at hand. The blockchain can be used in such a way that it has some positive benefits on the energy sector, but it is unlikely that it will create a massive restructuring of the energy sector in the short to medium term. As with most new technologies, some proponents of it believe that it will revolutionize the existing industry, and then, on the other hand, there are always critics of the technology who see it nothing more than a passing fad. Blockchain certainly has received significant attention from both sets of actors and the impact of the technology in the energy sector will fall somewhere between the two extreme camps.

6. Conclusion

This chapter has presented an overview of the potential impacts of the blockchain technology in the energy sector. This chapter has introduced the technology and the context surrounding not only the technology but also the ongoing energy transition. The combination of these two events has shown that the blockchain technology could play a significant role across numerous sectors in the future energy system.

This chapter has identified five areas within the energy sector where the blockchain technology could play a significant role. These applications were energy trading, environmental attribute management, demand response, electric mobility, and financing. Within the energy trading subsector, there are clear parallels between the blockchain technology and the operation of decentralized energy networks; however, there are still major challenges to overcome. Physical flows of electricity are incredibly hard to track, and it is not possible to prove the origin of the electricity consumed by an individual. The rise of distributed energy resources will cause the energy system to become more decentralized in the future, and electricity trading will also become more decentralized, and it is unclear just exactly what role the blockchain will play in decentralized energy trading.

With regard to blockchain's role in helping manage the environmental attributes of the various products in the energy sector, the role is more clearly defined, and the path to adoption may be easier than blockchain's use in energy trading. This is chiefly down to the fact that there is no need for a physical transfer of an asset within trading environmental attribute certificates. There is potential for the blockchain to speed the process of trading environmental certificates by removing some of the middlemen involved in the sector.

The two applications, demand response and electric mobility, could also benefit significantly from the blockchain technology, especially from the use of smart contracts to enable decentralized trading platforms for demand response and electric mobility, respectively. These platforms could open up these two fields to a number of small consumers or community energy initiatives by lowering the existing barriers to entry of the two markets.

The lowering of entry barriers and also the lowering of verification costs could also allow the blockchain technology to have a significant impact in the financing of energy projects. Tokenization of energy projects holds some interesting applications, and it could be used to finance various projects within the energy sector.

That being said, there are some major challenges that need to be overcome before the technology can realize its full potential. Not all of these challenges are related to the technical characteristics of the blockchain technology. Some of these challenges relate to the regulatory aspects of its deployment in the energy sector, and other challenges relate directly to the nature of the energy system.

Currently, there seem to be two schools of thinking relating to the blockchain's effect on the energy system. On the one hand, there are the blockchain advocates who believe that the technology will revolutionize the current operations of the energy system. On the other hand, there is a group who sees the technology as a passing fad with roots in illegal activities and get-rich-quick schemes. As the technology matures and society learns more about it and how to use it, it is hoped that these two camps can come closer together and work toward a middle ground where the positive effects of the technology are truly felt, not only in the energy system but also in the society as a whole. Proactive regulation is key to helping achieve this goal, and further research will key, especially using pilot projects, to investigate how the technology works in the real world.

Acknowledgment

M. Gough, S. Santos, and J.P.S. Catalão acknowledge the support by FEDER funds through COMPETE 2020 and by Portuguese funds through FCT, under POCI-01-0145-FEDER-029803 (02/SAICT/2017).

Appendix 1: Examples of companies using the blockchain technology in the energy sector

Blockchain companies working in the energy sector.

	Company	Sector	Location
1	4New	Cryptocurrencies, tokens, and investment	The United Kingdom
2	Aizu Laboratories	Grid management	Japan
3	Alastria	General purpose initiatives and consortia	Spain
4	Alliander and Spectral Energy (Jouliette at De Ceuvel)	Decentralized energy trading	Netherlands
5	Alliander (Alva)	Decentralized energy trading	Netherlands
6	Alliander (Charge Ledger)	Electric e-mobility	Netherlands
7	Assetron Energy	Cryptocurrencies, tokens, and investment	Australia
8	Bankymoon	Metering, billing, and security	South Africa
9	BAS Nederland	Metering, billing, and security	Netherlands
10	BCDC (Blockchain Development Company)	Cryptocurrencies, tokens, and investment	The United Kingdom
11	BCPG Group	Decentralized energy trading	Thailand
12	BittWatt	Decentralized energy trading	Romania

Blockchain companies working in the energy sector—cont'd

	Company	Sector	Location
13	BLOC (EnergyBlock and Community Power)	Decentralized energy trading	Denmark
14	Blockchain Futures Lab	General purpose initiatives and consortia	The United States
15	Blockchain Research Lab	General purpose initiatives and consortia	n/a
16	BlockLab	General purpose initiatives and consortia	Netherlands
17	Bouygues Immobilier and Stratumn	Decentralized energy trading	France
18	BTL	Decentralized energy trading	Canada and the United Kingdom
19	Car eWallet	Electric e-mobility	Germany
20	CarbonX	Green certificates and carbon trading	Canada
21	CGI and Eneco	Metering, billing, and security	Netherlands
22	Clearwatts	Decentralized energy trading	Netherlands
23	ClimateCoin	Green certificates and carbon trading	Switzerland
24	Conjoule	Decentralized energy trading	Germany
25	COSOL	Decentralized energy trading	Brazil
26	DAISEE	IoT, smart devices, automation, and asset management	France
27	Dajie	IoT, smart devices, automation, and asset management	The United Kingdom
28	DAO IPCI (MITO)	Green certificates and carbon trading	Russia
29	Department of Energy, The United States	Metering, billing, and security	The United States
30	Department of Energy, The United States	IoT, smart devices, automation, and asset management	The United States
31	Divvi	Decentralized energy trading	Australia
32	Dooak	Cryptocurrencies, tokens, and investment	Brazil
33	Drift	Decentralized energy trading	The United States
34	EcoCoin	Cryptocurrencies, tokens, and investment	Netherlands
35	Elbox	Decentralized energy trading	Switzerland

Continued

Blockchain companies working in the energy sector—cont'd

	Company	Sector	Location
36	ElectriCChain (SolarCoin)	IoT, smart devices, automation, and asset management	Andorra
37	Electrify.Asia	Decentralized energy trading	Singapore
38	Electron	Metering, billing, and security	The United Kingdom
39	Electron	Grid management	The United Kingdom
40	Elegant	Metering, billing, and security	Belgium
41	eMotorWerks	Electric e-mobility	The United States
42	Enbloc	Decentralized energy trading	The United States
43	Endesa Energia (Blockchain Lab)	General purpose initiatives and consortia	Spain
44	enercity	Metering, billing, and security	Germany
45	ENERES	Decentralized energy trading	Japan
46	EnergiMine	Cryptocurrencies, tokens, and investment	The United Kingdom
47	Energo Labs	Decentralized energy trading	China
48	Energo Labs	Electric e-mobility	China
49	Energy Bazaar	Decentralized energy trading	India
50	Energy Web Foundation	General purpose initiatives and consortia	Switzerland
51	Energy21 and Stedin	Decentralized energy trading	Netherlands
52	Energy-Blockchain Lab and IBM	Green certificates and carbon trading	China
53	EnerPort	Decentralized energy trading	Ireland
54	Enervalis (NRGCoin)	Cryptocurrencies, tokens, and investment	Belgium
55	Engie	Metering, billing, and security	France
56	EnLedger	Cryptocurrencies, tokens, and investment	The United States
57	Envion	Cryptocurrencies, tokens, and investment	Germany
58	EU Blockchain Observatory and Forum	General purpose initiatives and consortia	EU
59	Eurelectric (Blockchain Discussion Platform)	General purpose initiatives and consortia	EU
60	EverGreenCoin	Cryptocurrencies, tokens, and investment	The United States

Blockchain companies working in the energy sector—cont'd

	Company	Sector	Location
61	Everty	Electric e-mobility	Australia
62	Evolve Power	Grid management	The United States
63	Farad	Cryptocurrencies, tokens, and investment	UAE
64	Filament	Grid management	The United States
65	Filament	IoT, smart devices, automation, and asset management	The United States
66	Fortum	IoT, smart devices, automation, and asset management	Finland
67	Freeelio (AdptEVE)	IoT, smart devices, automation, and asset management	Germany
68	Green Energy Wallet	Cryptocurrencies, tokens, and investment	Germany
69	Green Running (Verv)	IoT, smart devices, automation, and asset management	The United Kingdom
70	Green Running (Verv)	Decentralized energy trading	The United Kingdom
71	Greeneum	Decentralized energy trading	Israel
72	Greeneum	Cryptocurrencies, tokens, and investment	Israel
73	Grid Singularity	Green certificates and carbon trading	Austria
74	Grid Singularity	Grid management	Austria
75	Grid+	Decentralized energy trading	The United States
76	Grünstromjeton	Cryptocurrencies, tokens, and investment	Germany
77	Hive Power	Decentralized energy trading	Switzerland
78	HydroMiner	Cryptocurrencies, tokens, and investment	Austria
79	IBM and Linux Foundation (Hyperledger)	General purpose initiatives, and consortia	The United States
80	ImpactPPA	Cryptocurrencies, tokens, and investment	The United States
81	Innogy MotionWerk (Share&Charge)	Electric e-mobility	Germany
82	Intrinsic ID and Guardtime	IoT, smart devices, automation, and asset management	The United States
83	Inuk	Cryptocurrencies, tokens, and investment	France

Continued

Blockchain companies working in the energy sector—cont'd

	Company	Sector	Location
84	KEPCO	Decentralized energy trading	Japan
85	LO3 Energy	Decentralized energy trading	The United States
86	Local-e	Cryptocurrencies, tokens, and investment	The United States
87	Marubeni (Coincheck Denki)	Metering, billing, and security	Japan
88	M-PAYG	Metering, billing, and security	Denmark
89	MyBit	Cryptocurrencies, tokens, and investment	Switzerland
90	Nasdaq New York Linq	Green certificates and carbon trading	The United States
91	OLI	IoT, smart devices, automation, and asset management	Germany
92	Omega Grid	Decentralized energy trading	The United States
93	OneUp	Decentralized energy trading	Netherlands
94	OurPower (CEDISON)	Grid management	The United Kingdom
95	Oursolargrid & ITP	Decentralized energy trading	Germany
96	Oxygen Initiative	Electric e-mobility	The United States
97	PetroBloq	Decentralized energy trading	Canada
98	Platinum Energy Recovery	Decentralized energy trading	Singapore
99	PONTON (EnerChain)	Decentralized energy trading	Germany
100	PONTON (GridChain)	Grid management	Germany
101	Poseidon	Green certificates and carbon trading	Switzerland
102	Power Ledger (EcoChain)	Decentralized energy trading	Australia
103	Power Ledger	IoT, smart devices, automation, and asset management	Australia
104	Power Ledger	Electric e-mobility	Australia
105	Power Ledger	Green certificates and carbon trading	Australia
106	Power Ledger	Grid management	Australia
107	Power-ID	Decentralized energy trading	Switzerland
108	PROSUME	Metering, billing, and security	Switzerland

Blockchain companies working in the energy sector—cont'd

	Company	Sector	Location
109	PROSUME	Cryptocurrencies, tokens, and investment	Switzerland
110	PROSUME	Decentralized energy trading	Switzerland
111	PROSUME	Grid management	Switzerland
112	PROSUME	Electric e-mobility	Switzerland
113	PRTI	Cryptocurrencies, tokens, and investment	The United States
114	Pylon Network	Metering, billing, and security	Spain
115	Pylon Network	Decentralized energy trading	Spain
116	Restart Energy	Decentralized energy trading	Romania
117	Slock.it	IoT, smart devices, automation, and asset management	Germany
118	Slock.it	Electric e-mobility	Germany
119	Solar bankers (SunCoin)	Decentralized energy trading	Singapore
120	Solar DAO	Cryptocurrencies, tokens, and investment	Israel
121	SolarChange (SolarCoin)	Cryptocurrencies, tokens, and investment	Andorra
122	SP Energy Networks, SSEN, SP Distribution, SP Manweb, and UK Power Networks	Grid management	The United Kingdom
123	Spectral Energy	Decentralized energy trading	Netherlands
124	STROMDAO	Decentralized energy trading	Germany
125	SunChain (TECSOL and Enedis)	Metering, billing, and security	France
126	SunContract	Decentralized energy trading	Slovenia
127	Swytch	IoT, smart devices, automation, and asset management	South Korea
128	Tavrida Electric	IoT, smart devices, automation, and asset management	Russia
129	TenneT and sonnen	Grid management	Netherlands
130	TenneT and Vandenbron	Grid management	Netherlands
131	The Sun Exchange	Cryptocurrencies, tokens, and investment	South Africa
132	TOBLOCKCHAIN	Decentralized energy trading	Netherlands
133	toomuch.energy	Decentralized energy trading	Belgium

Continued

Blockchain companies working in the energy sector—cont'd

	Company	Sector	Location
134	ubitricity	Electric e-mobility	Germany
135	VAKT and partners (including BP, Shell, and Statoil)	Decentralized energy trading	The United Kingdom
136	Vattenfall (Powerpeers)	Decentralized energy trading	Netherlands
137	Vector Energy (EcoChain)	Decentralized energy trading	New Zealand
138	Veridium Labs	Green certificates and carbon trading	Hong Kong
139	Volt Markets	Decentralized energy trading	The United States
140	Volt Markets	Green certificates and carbon trading	The United States
141	Wanxiang	IoT, smart devices, automation, and asset management	China
142	WePower	Cryptocurrencies, tokens, and investment	Gibraltar
143	Wien Energie	Decentralized energy trading	Austria
144	Wirepas	IoT, smart devices, automation, and asset management	Finland
145	Wuppertal Stadtwerke (Tal.Markt)	Decentralized energy trading	Germany
146	XinFin	Cryptocurrencies, tokens, and investment	Singapore
147	XiWATT	Cryptocurrencies, tokens, and investment	The United States

Authors own after M. Andoni, V. Robu, D. Flynn, S. Abram, D. Geach, D. Jenkins, P. McCallum, A. Peacock, Blockchain technology in the energy sector: a systematic review of challenges and opportunities, Renew. Sustain. Energy Rev. 100 (2019) 143–174. https://doi.org/10.1016/j.rser.2018. 10.014 and D. Livingston, V. Sivaram, M. Freeman, M. Fiege, Applying Blockchain Technology to Electric Power Systems, 2018.

References

[1] S.P. Burger, M. Luke, Business models for distributed energy resources: a review and empirical analysis, Energy Policy 109 (2017) 230–248, https://doi.org/10.1016/j. enpol.2017.07.007.

[2] C. Burger, A. Kuhlmann, P. Richard, J. Weinmann, Blockchain in the Energy Transition. A Survey Among Decision-Makers in the German Energy Industry | ESMT Berlin, Deutsche Energie-Agentur GmbH (dena), Berlin, Germany, 2016.

[3] European Commission, Monitoring Progress Towards the Energy Union Objectives – Key Indicators (Working Paper), European Commission, Brussels, Belgium, 2017.

[4] D. Livingston, V. Sivaram, M. Freeman, M. Fiege, Applying Blockchain Technology to Electric Power Systems, 2018.

[5] C. Catalini, J.S. Gans, Some Simple Economics of the Blockchain (SSRN Scholarly Paper No. ID 2874598), Social Science Research Network, Rochester, NY, 2017.

[6] IBM, Device Democracy Saving the Future of the Internet of Things, 2015.

[7] M. Luke, S. Lee, Z. Pekarek, A. Dimitrova, Blockchain in Electricity: A Critical Review of Progress to Date, 2018.

[8] P. Vigna, M. Casey, The Truth Machine: The Blockcahin and the Future of Everything, St. Martin's Press, 2018.

[9] J. Mattila, The Blockchain Phenomenon – The Disruptive Potential of Distributed Consensus Architectures (No. 38), ETLA Working Papers, The Research Institute of the Finnish Economy, 2016.

[10] M.A. Jamison, P. Tariq, Five things regulators should know about blockchain (and three myths to forget), Electr. J. 31 (2018) 20–23, https://doi.org/10.1016/j.tej.2018.10.003.

[11] S. Meunier, Chapter 3: Blockchain 101: what is blockchain and how does this revolutionary technology work? in: A. Marke (Ed.), Transforming Climate Finance and Green Investment with Blockchains, Academic Press, 2018, pp. 23–34. https://doi.org/10.1016/B978-0-12-814447-3.00003-3.

[12] R. Chitchyan, J. Murkin, Review of Blockchain Technology and its Expectations: Case of the Energy Sector (Working Paper), University of Bristol, Department of Computer Science, 2018.

[13] A. Woodhall, How Blockchain can democratize global energy supply, in: Transforming Climate Finance and Green Investment with Blockchains, Elsevier, 2018.

[14] I. Bashir, Mastering Blockchain: Distributed Ledger Technology, Decentralization, and Smart Contracts Explained, second ed., Packt Publishing Ltd., 2018.

[15] M. Andoni, V. Robu, D. Flynn, S. Abram, D. Geach, D. Jenkins, P. McCallum, A. Peacock, Blockchain technology in the energy sector: a systematic review of challenges and opportunities, Renew. Sustain. Energy Rev. 100 (2019) 143–174, https://doi.org/10.1016/j.rser.2018.10.014.

[16] J.A.F. Castellanos, D. Coll-Mayor, J.A. Notholt, Cryptocurrency as guarantees of origin: Simulating a green certificate market with the Ethereum Blockchain, in: 2017 IEEE International Conference on Smart Energy Grid Engineering (SEGE), in: Presented at the 2017 IEEE International Conference on Smart Energy Grid Engineering (SEGE), 2017, pp. 367–372. https://doi.org/10.1109/SEGE.2017.8052827.

[17] D. Shipworth, Peer-to-Peer Energy Trading Using Blockchains, 2017.

[18] S. Nakamoto, Bitcoin: A Peer-to-Peer Electronic Cash System, 2008.

[19] Go Bitcoin, Cost of a 51% attack – GoBitcoin [WWW Document], GoBitcoin.io, 2019. https://gobitcoin.io/tools/cost-51-attack/. (Accessed 15 January 2019).

[20] S. Muftic, I. Sanchez, L. Beslay, Overview and Analysis of the Concept and Applications of Virtual Currencies (No. EUR 28386 EN), Italy, Joint Research Centre, European Commission, Ispra, 2016.

[21] D. Romano, G. Schmid, Beyond bitcoin: a critical look at blockchain-based systems, Cryptography (2017) 1, https://doi.org/10.3390/cryptography1020015.

[22] M.L.D. Silvestre, P. Gallo, M.G. Ippolito, E.R. Sanseverino, G. Sciumè, G. Zizzo, An energy blockchain, a use case on tendermint, in: 2018 IEEE International Conference on Environment and Electrical Engineering and 2018 IEEE Industrial and Commercial Power Systems Europe (EEEIC/I CPS Europe), in: Presented at the 2018 IEEE International Conference on Environment and Electrical Engineering and 2018 IEEE Industrial and Commercial Power Systems Europe (EEEIC/I CPS Europe), 2018, pp. 1–5. https://doi.org/10.1109/EEEIC.2018.8493919.

[23] M. Mihaylov, I. Razo-Zapata, A. Nowe, NRGcoin—a blockchain-based reward mechanism for both production and consumption of renewable energy, in: Transforming Climate Finance and Green Investment with Blockchains, Elsevier, 2018, p. 368.

[24] N. Tomaino, Cryptoeconomics 101 [WWW Document], The Control, 2017. https://thecontrol. co/cryptoeconomics-101-e5c883e9a8ff. (Accessed 8 January 2019).

[25] E. Mengelkamp, J. Gärttner, K. Rock, S. Kessler, L. Orsini, C. Weinhardt, Designing micro-grid energy markets: a case study: the Brooklyn Microgrid, Appl. Energy 210 (2018) 870–880, https://doi.org/10.1016/j.apenergy.2017.06.054.

[26] P. Verma, B. O'Regan, B. Hayes, S. Thakur, J.G. Breslin, EnerPort: Irish Blockchain project for peer-to-peer energy trading, Energy Inform. 1 (2018), https://doi.org/10.1186/s42162-018-0057-8.

[27] Energy Web Foundation, n.d. Energy Web Foundation [WWW Document]. https:// energyweb.org/ (Accessed 10 January 2019).

[28] Hyperledger, Blockchain technology projects, Hyperledger, 2019. https://www.hyperledger. org/projects. (Accessed 10 January 2019).

[29] T. Lyons, Blockchain Innovation in Europe, EU Blockchain Observatory and Forum, Brussels, Belgium, 2018.

[30] P. Brody, How blockchains will industrialize a renewable grid, in: Transforming Climate Finance and Green Investment with Blockchains, Elsevier, 2018.

[31] Blockchain.com, Blockchain Explorer | BTC | ETH [WWW Document], https://www. blockchain.com/explorer, 2019. (Accessed 10 January 2019).

[32] D. Larimer, EOSIO Dawn 3.0 Now Available, eosio, 2018. https://medium.com/eosio/eosio-dawn-3-0-now-available-49a3b99242d7. (Accessed 15 January 2019).

[33] Bloks.io, Bloks.io | Fastest EOS Block Explorer and Wallet – EOS Cafe & HKEOS [WWW Document], https://bloks.io, 2019. (Accessed 15 January 2019).

[34] Open Trading Network, How Crypto-Kitties Disrupted the Ethereum Network [WWW Document], Hacker Noon, 2017. https://hackernoon.com/how-crypto-kitties-disrupted-the-ethereum-network-845c22aa1e6e. (Accessed 10 January 2019).

[35] M. Orcutt, Ethereum thinks it can change the world. It's running out of time to prove it. [WWW Document], MIT Technology Review, 2018. https://www.technologyreview.com/s/612507/ ethereum-thinks-it-can-change-the-world-its-running-out-of-time-to-prove-it/. (Accessed 13 December 2018).

[36] Competition & Markets Authority, Energy Market Investigation, Competition & Markets Authority, London, 2016.

[37] D. Pollock, How Can Blockchain Thrive In The Face Of European GDPR Blockade? [WWW Document], Forbes, 2018. https://www.forbes.com/sites/darrynpollock/2018/10/03/how-can-blockchain-thrive-in-the-face-of-european-gdpr-blockade/. (Accessed 11 January 2019).

[38] K. Rooney, SEC Chairman Clayton says agency won't change definition of a security [WWW Document], https://www.cnbc.com/2018/06/06/sec-chairman-clayton-says-agency-wont-change-definition-of-a-security.html, 2018. (Accessed 11 January 2019).

[39] J. Hwang, M. Choi, T. Lee, S. Jeon, S. Kim, S. Park, S. Park, Energy prosumer business model using blockchain system to ensure transparency and safety, Energy Procedia 141 (2017) 194–198, https://doi.org/10.1016/j.egypro.2017.11.037.

[40] M. Merz, Potential of the blockchain technology in energy trading, in: Blockchain Technology Introduction for Business and IT Managers, De Gruyter, Berlin, Germany, 2016.

[41] C.G. Monyei, K.E.H. Jenkins, Electrons have no identity: setting right misrepresentations in Google and Apple's clean energy purchasing, Energy Res. Soc. Sci. 46 (2018) 48–51, https:// doi.org/10.1016/j.erss.2018.06.015.

[42] C. Park, T. Yong, Comparative review and discussion on P2P electricity trading, in: Energy Procedia, International Scientific Conference "Environmental and Climate Technologies",

CONECT 2017, 10–12 May 2017, Riga, Latvia 128, 3–9, 2017. https://doi.org/10.1016/j.egypro.2017.09.003.

[43] S. Noor, W. Yang, M. Guo, K.H. van Dam, X. Wang, Energy demand side management within micro-grid networks enhanced by blockchain, Appl. Energy 228 (2018) 1385–1398, https://doi.org/10.1016/j.apenergy.2018.07.012.

[44] M. Hinterstocker, C. Dufter, S. Roon, A. von Bogensperger, A. Zeiselmair, Potential impact of Blockchain solutions on energy markets, in: 2018 15th International Conference on the European Energy Market (EEM), in: Presented at the 2018 15th International Conference on the European Energy Market (EEM), 2018, pp. 1–5. https://doi.org/10.1109/EEM.2018.8469988.

[45] A. Galenovich, S. Lonshakov, A. Shadrin, Blockchain ecosystem for carbon markets, environmental assets, rights, and liabilities: concept design and implementation, in: Transforming Climate Finance and Green Investment with Blockchains, Elsevier, 2018.

[46] South Pole, SouthPole, ixo Foundation, GoldStandard develop blockchain application for carbon credit tokenization [WWW Document], South Pole, 2018. https://www.southpole.com/media-corner/southpole-ixo-goldstandard-blockchain-application-for-carbon-credit-tokenization. (Accessed 14 January 2019).

[47] D. Sanders, A. Hart, M. Ravishankar, J. Brunert, G. Strbac, M. Aunedi, D. Pudjianto, An analysis of electricity system flexibility for Great Britain, Carbon Trust, London, 2016.

[48] D. Martin, Blockchain in the Marine Bunker Market [WWW Document], gCaptain, 2018. https://gcaptain.com/blockchain-in-the-marine-bunker-market/. (Accessed 18 February 2019).

Chapter 3

Transition toward blockchain-based electricity trading markets

Mohamed Lotfi[a,b], Cláudio Monteiro[a], Miadreza Shafie-khah[c] and João P.S. Catalão[a,b]

[a]*Faculty of Engineering, University of Porto, Porto, Portugal*, [b]*INESC TEC, Porto, Portugal*, [c]*School of Technology and Innovations, University of Vaasa, Vaasa, Finland*

1. The rise of blockchain

It may be surprising to recall that blockchain was unheard of merely a decade ago. Being first launched in 2008 in the advent of Bitcoin, it is indeed astounding how the technology rapidly became a major enabler of all sorts of decentralized platforms, which are now ever so important in this digital era of extensive Internet-of-Thing (IoT) enabling. In order to understand how blockchain-based systems took center stage in modern electricity trading frameworks, it is necessary to look back at the sequence of events which resulted in the need for blockchain, its creation, success as the first fully decentralized commercial platform, and subsequent expansion to the energy sector.

In fact, blockchain seems to have emerged at the most convenient timing for the energy sector, which in itself was transforming in favor of more decentralized structures and decision making. Therefore, blockchain appeared as a reliable solution to several challenges facing the energy sector, and with perfect timing. That being said, this convenience is absolutely no coincidence. These simultaneous events happened, and continue to happen, in the context of a much larger revolution, the fourth industrial revolution.

1.1 The fourth industrial revolution

Previous industrial revolutions all had one thing in common: each of them was triggered by a uniquely identifiable technological breakthrough. Steam engines fueled mechanization in the first, electricity sparked mass production in the second, and electronics and computers made automation possible in the third.

Blockchain-based Smart Grids. https://doi.org/10.1016/B978-0-12-817862-1.00003-8

TABLE 1 Number of IoT-connected devices and a summary of predicted changes from 2016 to 2021.

Type of device	IoT connected in 2016 (billion)	IoT connected in 2021 (billion)	Percentage change
All	17.6	28 (Ericsson) [8] 30.7 (IHS) [9]	+60% ⇔ +75%
Conventional (computers, smartphones, etc.)	11.2	7.2 (Ericsson/ Gartner) 2.6 (IHS/IDC)	−35% ⇔ −75%
"Things" (appliances, sensors, etc.)	6.4	20.8 (Gartner) [7] 28.1 (IDC) [10]	+225% ⇔ +340%

This rapid uncontrollable growth of IoT is unleashing unprecedented data traffic on the Internet, creating two main problems: *data redundancy* and *data security/privacy*. In an IoT stakeholders survey [11], 41% of respondents said "timely collection and analysis of data" was a major challenge since there was "too much data to analyze effectively," "difficult to capture useful data," and "data is analyzed too slowly to be actionable." Those responses precisely describe the data redundancy problem in modern IoT-enabled systems.

A major solution effort to data redundancy was the development cloud computing. IoT refers to the connection of devices to the Internet and cloud computing refers to how those devices work together to deliver data, applications, or services [12]. IBM defines cloud computing as the "delivery of on-demand computing resources... over the Internet on a pay-for-use basis" [13]. Another effort to tackle the data redundancy problem has been the development of more advanced and efficient distributed data analysis algorithms.

In fact, it is interesting to see that according to all forecasts the number of conventional devices is expected to decrease, despite the enormous overall IoT growth. This can be attributed to the fact that with the increased use of things like smart sensors or smart actuators, the need for many computers currently used solely to provide the Internet link for such devices will cease to exist. In addition, the emergence of cloud computing and more advanced data analytics will facilitate shared processing and storage resources, reducing the required number of dispensable computing and storage devices.

As for the second (security/privacy) problem, it is hardly possible to come across any IoT-related discussion without the mention of the topic. With everything from personal appliances to industrial machinery being connected to an extended global network, the potential damage of cyberattacks and unsolicited disclosure could be devastating. It is therefore not surprising that another report by Gartner [14] predicted that IoT security spending growth is set to overtake overall IoT spending growth by 2017, which was an impressively accurate prediction. In the survey of IoT stakeholders mentioned earlier, the top challenge in IoT projects reported by 58% of respondents was "Business processes or policies" in which they complained that privacy concerns over confidential data posed a major issue preventing data collection [11]. The approval of the European Union's General Data Protection Regulation (GDPR) in 2016 [15] and other similar legislation worldwide made data security and privacy not only a concern, but also a legally binding obligation for all sectors affected by IoT enabling.

Those sectors were therefore expected to explore and implement novel solutions to address data redundancy and security issues. While the issue of data redundancy was quickly handled early on, privacy and security issues remained a major concern. This was due to two reasons. One of the main reasons for this is the fact that the latter is not only dependent on the availability of feasible technical solutions, but also involves social, political, and economical debate.

The energy industry is and will continue to be one of the most affected by the growth of IoT. Of the 20.8 billion nonconventional devices expected to be online by 2020/2021, around 1.4 billion will be from the energy industry, and 1.5 billion from home energy management devices. This meant that 10% of all IoT endpoints will be energy or energy management devices. Therefore, the energy sector began to realize in that period that incorporating compatible and feasible solutions for both data redundancy and privacy problems will be a necessity for the design future energy system structures in an IoT-dominated world.

This unstoppable IoT enabling of energy systems on all levels reinforced an increasingly popular vision in scientific literature: the Internet of Energy (IoE). This vision of IoE being the product of IoT enabling of smart grids (SGs) was first mentioned in 2010 [16]. The article envisioned scalable and self-sufficient energy networks through Internet enabling. Computational power required for coordination and management of energy supply and demand is provided by cloud resources. The other stated requirement was sufficient energy storage resources, which is becoming increasing efficient and affordable. As such, scientific literature started showing great attention to this IoE paradigm [17–23] with multiple other associated variants such as Local Area Energy Networks (E-LAN) [24] and Smart Grids 2.0 [25].

The consensus in scientific literature was that technical models and processes developed in an IoE paradigm should be: (1) distributed (fully decentralized), (2) efficient at data analysis (with efficient forecasting and optimization capabilities), (3) scalable, and (4) user-friendly (plug and play) [20, 23]. Those correspond exactly with design requirements listed earlier, making this IoE framework a perfectly suitable as an I4.0 solution model.

1.3 Decentralized economies: The success of blockchain

Following the 2008 global financial crisis, the world's first digital cryptocurrency (Bitcoin) was proposed [26]. The introduced platform allowed peer-to-peer (P2P) transactions to take place, eliminating the need for intermediary financial authorities, being the first fully decentralized commercial system of its kind. It was a tremendous success. In 2010, 1 Bitcoin was valued at 0.08 USD, and rose exponentially to reach a peak value of 17,000 USD in 2017, maintaining a market capital well above 100 Billion USD since then. This astonishingly rapid success is primarily attributed to the underlying technology: blockchain, a cryptographically secured distributed database containing blocks of transactions.

The platform possesses two distinguishing characteristics allowing it to provide a decentralized system: security and global consensus. The latter is provided by the fact that everyone in the network is constantly validating and updating the state of the system collectively. Since each block in the chain is linked to the previous one, all users can verify if contents have not been modified. Keys are function of both the encrypted contents of the block and the previous block's key, thus involve a puzzle to be solved requiring computational effort. Keys are generated by miners: users providing the distributed computational effort and rewarded accordingly. The platform's decentralized nature makes it immune to many cyberattacks, even if a large number of users are targeted.

With the conception of Ethereum in 2012 and its launch in 2015, Blockchain 2.0 introduced smart contracts: digitally written and signed awaiting satisfaction of certain conditions to come into effect [27]. Ethereum's Blockchain 2.0 with its smart contracts was an equally massive success, amassing a market capital of over 1 Billion USD within less than a year of its launch, which exponentially grow to steadily remain above 10 Billion USD since 2017 (Fig. 2).

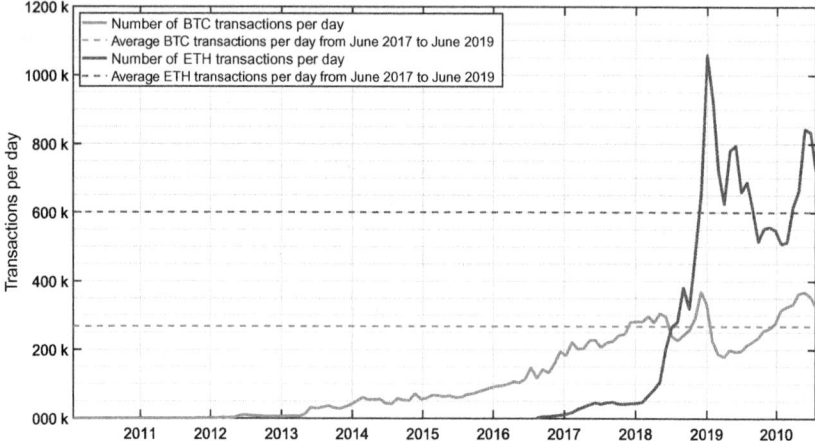

FIG. 2 The phenomenal success of blockchain technology is shown by the number of daily transactions taking place on Bitcoin and Ethereum, currently the world's two largest cryptocurrencies.

With hundreds of billions of market capital, blockchain-based trading platforms have clearly gained society's trust, which became impossible to miss. Official recognition of cryptocurrencies and their underlying technology was inevitable. In 2012, the European Central Bank first recognized digital currencies [28] and later in 2015 just before Ethereum was about to be launched, a follow-up report was released with an extensive analysis of the success of their decentralized platforms [29]. This consolidated the acknowledgement of blockchain-based systems. During the past 2 years (between June 2017 and June 2019), there has been an average of more than 250,000 daily Bitcoin transactions, and more than 600,000 Ethereum daily transactions. Bitcoin and Ethereum continue to dominate as the two leading digital currencies with market capitals well above 10 and 100 Billion USD since 2017, respectively. However, there are numerous other blockchain-based cryptocurrencies which have emerged, with hundreds of thousands of daily transactions

The massive success and recognition of blockchain with its smart contracts as a decentralized commercial system has led many people to investigate the application in different sectors, particularly those that are shifting most toward a decentralized structures. The secure cryptographic algorithm of blockchain and its immunity to many cyberattacks is even more reason why it is currently seen as an enabling technology as it may potentially offer a solution to many data security/privacy problems caused by IoT enabling.

2. Adoption of blockchain by the energy sector

Around the same time when blockchain-based commercial platforms were rising and gaining global recognition, the energy sector was going through a massive transformation of its own. Motivated by a triad of causes (security of supply, environmental protection, and economic efficiency), legislation was being passed worldwide eagerly promoting demand-side management (DSM) strategies, specifically demand response (DR) programs. DR inherently relies on the availability of two main things: distributed energy resources (DERs) and SG infrastructure (with smart metering and communication devices). In 2012, the EU passed a directive to direct the rollout of SGs to implement DR programs (and multiple similar legislation was passed worldwide contemporarily) [30]. This ultimately meant that power and energy systems were about to rapidly witness two major transformations: physical decentralization due to DER installations and information decentralization due to smart metering and SG rollout.

It is important to elaborate that decentralization can occur at three different and distinguishable layers:

Decentralization of power systems: This is related to the physical disaggregation of power generation, for example, DERs.

Decentralization of information systems: Due to interconnection of even the smallest devices as in the IoT paradigm.

Decentralization of energy markets: Is the case with P2P trading of generated energy by prosumers.

An advantage of decentralized systems is their capacity to make better use of the local endogenous resources and reduce costs and losses of transporting these resources, consequently leading to more environmental sustainability. Economic sustainability depends on the scale factor; big centralized systems are more efficient with low unitary costs. Small decentralized systems are less efficient with high unitary costs. However, recent technological developments of decentralized systems are enabling them to be economically competitive with their centralized counterparts. The three levels of decentralization are interinfluential and complementary.

*Between P2P energy trading and IoE there is a clear emergence of what is referred to as **democratic energy systems** in which fundamental aspects are (1) significant citizen participation and (2) decentralized decision making in operation, management, planning, and trading, and (3) RES-dominated generation.*

2.1   P2P energy markets: The emerging paradigm shift

Conventional electrical power systems had unidirectional energy flow from generation to consumption. A centralized structure was best suited for this model with different utilities managing operation, planning, and energy market operations. Increased penetration of prosumers with DERs made electric power systems more decentralized and rendering the conventional model obsolete. First, energy was now being generated at both ends of the conventional chain and therefore roles of operators and utilities need to be redefined or the structure shuffled altogether. Second, DERs are increasingly incorporated into energy networks without being given any operational role or access to the wholesale market which is not sustainable [31].

In the beginning, feed-in-tariffs were offered with the intention of incentivizing small consumers to install small renewable energy generation (e.g., rooftop solar PV). Consumers generating electricity with renewable sources would be able to feed any surplus energy into the grid and are paid for it, albeit at a rate which is significantly lower than the electricity market price. This among other reasons started creating distrust between large utilities and system operators on one hand and DER owners on the other. With the decreasing price of renewable installations and the ease of acquiring them, small prosumers start looking for alternatives of trading electricity which can eliminate need for a middle man such as P2P trading.

After witnessing its capability to provide fully decentralized commercial trading platforms, it was clear that blockchain offered the ideal solution for newly emerged prosumers and their desire for citizen-run democratic energy systems. Multiple successful tests of blockchain-based P2P platforms started being carried out, albeit on small scales. Prior to the launch of Ethereum's Blockchain 2.0 and smart contracts in 2015, the role of blockchain was limited to enabling a secure and reliable distributed ledger of transactions, and thereby the early experiment with blockchain in P2P energy trading were strictly limited to its use to record financial transactions.

2012 was an important year in the transition to blockchain-based applications in the energy sector. The EU directive for SG rollout was approved, incentivizing researchers and stakeholders to seek new innovative data models and manage these new smart interconnected microgrids with DERs [30]. The first academic article putting forth the concept of "transactive energy" was published in 2013, proposing a vision of decentralized and self-sustaining microgrids capable of autonomous transactive operation [32].

This sparked a new trend in academic research, attempting and designing solutions based on this transactive energy vision. It is important to recall that at the time of the first transactive energy publication (early 2013), only first generation blockchain platforms have been in operation. Without smart contracts, and being limited only to financial transactions, the potential of blockchain being a suitable enabler for such a system was extremely limited. Thus, in the early years of research on transactive energy, blockchain was seldom mentioned.

With the proposal of Blockchain 2.0 in late 2012 and the launch of Ethereum in late 2015, successfully incorporating smart contracts, this second generation of blockchain technology was suitable to provide for the needs of the energy sector. Smart contracts made energy trading possible in the way that was being envisioned by researchers on transactive energy networks. Therefore, a few months after the successful launch of Ethereum and the witness of its success, the first academic research papers proposing a validated methodology for blockchain applications to energy systems and peer-to-peer electricity trading were made toward the end of 2015 and the beginning of 2016 (Fig. 3) [33–35].

2.2   Expansion of blockchain applications

Once the first proposals were presented for the application of blockchain in the energy sector, its expansion became exponential. In 2016, recognition of blockchain as an inevitable enabler of future energy grids and markets became obvious around the world. The major reports were published in 2016 by global consultancy firms and governmental agencies which investigated the status-quo of blockchain applications at the time and predicated its great potential in the years to come.

PricewaterhouseCoopers (PwC) released a report [36] in highlighting the opportunities blockchain offers for energy producers and consumers. The report started by stating that blockchain's transaction model which shifts from centralized structures to P2P can reduce costs and speed up processes resulting in more

Major milestones in the transition to blockchain-based energy trading

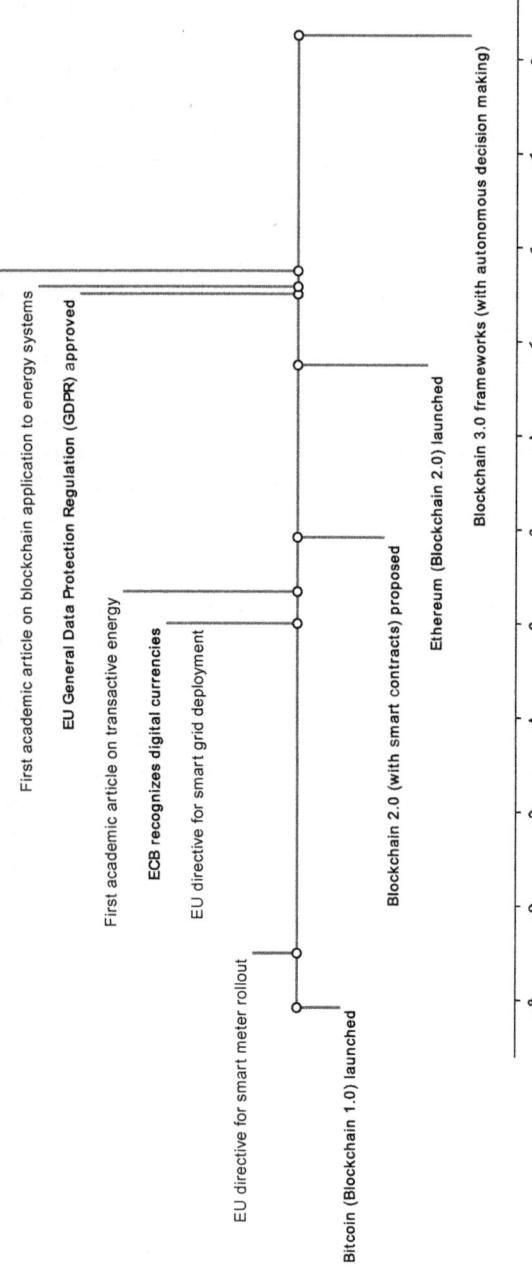

FIG. 3 Timeline of major milestones and milestones taking place in the transition toward blockchain-based energy trading.

This makes it an enabler technology for platforms with fully decentralized control which is particularly useful in situations where transacting parties lack trust. In addition, blockchain offers more efficient structures by removing the need for data to be synchronized with and by an intermediary, which is particularly useful in industry-level applications. For global and cross-country applications, the main potential of blockchain technology lied in its ability to offer interoperability between devices and systems. As such, the report identified that interoperability and flexibility is the target state which blockchain development in the energy sector must aim for. A conceptual use case which involved using blockchain in conjunction with smart meters was made and evaluated with industry specialists. The report recommended that being a disruptive technology to the energy sector, companies planning on developing blockchain energy applications should build strategic understanding of use cases in collaboration with blockchain technology developers by insourcing the knowledge.

An article [35] published in 2016 modeled and simulated a case of P2P energy trading in an SG. The proposed model was a fully decentralized private trading platform based on blockchain with multisignatures and anonymous encrypted message propagation streams. The system was resistant against significant common cyberattacks. In addition, the privacy of trading parties was found to be well protected by the system. By comparison with the results with a simulated centralized system, it concluded that the proposed system is a feasible application of blockchain technology to develop secure and efficient fully decentralized energy trading platforms.

A whitepaper [39] published by ParisTech in 2016 investigated new models for managing distribution grids. They proposed creating virtual distribution grids as a layer above the physical one. A blockchain-based platform would be used for transactions of surplus energy from homes in a distributed architecture. The proposed model was neither implemented nor tested. However, the paper attempted to provide blockchain usage model in the energy industry which is compatible with the current market structures. A journal [40] published in 2017 studied the energy market in Perth, Australia, where a recent successful experiment with a blockchain-trading platform was performed (similar Brooklyn). It showed that proliferation of cheap renewable generation and battery storage technologies are going to soon result in a paradigm shift in the energy industry to what they referred to as "citizen utilities." The paper states that an inevitable shift to distribute and bidirectional energy systems and more decentralized energy markets will take place, where blockchain will be the basis of transactions such systems. The response of traditional market players would be what the paper called a "fight, flight, or innovate" one: fight will be the case if markets are resistant and in denial of the new paradigm shift; flight is if energy utilities take no action and possibly divesting investment in traditional markets; and innovate is if current utilities embrace the new technologies driving the paradigm shift.

flexible systems. The report highlights that while some level of ⌐
being reached in the financial sector, the technology is still being
for other applications with some barriers in the way, primarily confl
tion and legal and regulatory requirements for fully decentralized sy:
Brooklyn Microgrid project was a successful experiment in which a
consisting of a group of 10 households directly traded surplus solar e1
erated using a blockchain system. Smart meters were used in conjun
blockchain's smart contracts to keep track of energy produced and t
ically effect transactions, respectively. An energy token system wa
energy payments. Most start-ups working on blockchain applications ;
were developing cryptocurrencies specific for energy trading. T
emphasized the opportunity blockchain offers for prosumers of ele
a P2P system, by providing more flexible and autonomous systems. I1
the report highlights that blockchain could potentially be employed
range of uses other than energy transactions, which include docu1
of ownership (of energy generated), guarantees of origin, renewable e1
tificates, and others. While blockchain could radically transform the e1
tor, the report stated that current legal and regulatory frameworks n
adjusted to cope with large-scale decentralized transactions models to
possible.

The German Energy Agency (dena) conducted a survey [37
70 decision makers in the German energy industry regarding bl
applications; 69% said they had already heard of existing blockchain
tions in the energy sector and 52% either have blockchain implement
ongoing plans thereof; 81% of the respondents are confident that bl
will likely have a significant influence on the industry. Potential use c
they envision were (in decreasing order of potential): security, decer
generation, P2P trading, mobility, metering and data transfer, tradi
forms, automation, billing, grid management, and communication.
chain's potential in cost reduction and as an enabler for new |
models was reported. Since it was expected to be more disruptive c
to current technological alternatives, it had a higher chance of being t
inant design in applications where P2P trading has not yet been establi
a large scale. Despite changing the structure of energy trading, if blo
applications prove to have monetary or timely advantages over existi1
tions, the critical number of market participants would be convinced t
don current platforms in favor of blockchain. Rapid successful launc
prototypes around the world might make Germany and the EU lagging
globally with current regulator frameworks being completely unsuital
uncompliant with blockchain applications. Thus, they urged policyma
consider it as a top priority.

Another report [38] studied the development of blockchain use ca
assuming the role of R&D developers. First, global consensus was identi
the primary disruptive element of blockchain technology in the energy

This chapter concludes that there is a rapid change to a new energy market model which is operated not by utilities, but by consumers and that this should be facilitated in what the author called "democratization of power."

3. Blockchain 3.0: Next-generation energy systems

Up to this stage, all published works and conducted experiments considered the capabilities of Blockchain 2.0. The cryptographically secured, consensus-based, approach enabled the elimination of financial mediators. Similarly, smart contracts enabled a fully decentralized market were energy can be traded. However, there was still a major pillar missing from the IoE vision for a fully autonomous transactive energy network. As mentioned earlier, there are three distinguishable layers of energy systems: the grid, the markets, and the information infrastructure. While Blockchain 2.0 solutions provided a way of managing the latter two in a fully decentralized fashion, it was not sufficient to be applied on the first. Operation and control of power systems require the solution of complex optimization and forecasting models, and it was still extremely challenging at that stage to develop a fully decentralized operation or control framework for electrical grids which would justify the dispensability of a (central) grid operator.

Only one academic paper at the time was proposed a blockchain-based solution for distributed optimization and control of electric grids in a P2P market architecture [41]. A decentralized optimal power flow (OPF) model for scheduling DERs on a microgrid was built and tested. Distributed optimization (namely ADMM) was used to decompose the OPF problem making it compatible with blockchain architecture. The cost function was decomposed into a set of local functions, and a global function which is a function of the local ones. In this manner, the scheduling and dispatch routine could be performed in a fully decentralized fashion using blockchain and smart contracts. The model was tested on a 55-bus microgrid with a dispatchable central generator, uncontrolled plug loads, nondispatchable renewable energy sources, shapeable loads, deferrable loads, and batteries. A day-ahead scheduling problem was considered with 1-h intervals. Blockchain and smart contracts used to perform optimization and control actions, and clearing prices, recording energy consumption (smart meters), and billing contracts (payment, charges, and penalties). The optimal cost based on ADMM was 0.4% larger than the centralized one. Shorter time horizons, ancillary services, or stochastic behavior were not considered. The aim was providing proof of the feasibility of using blockchain for distributed optimization and control grid applications. The success was due to the combination of Blockchain 2.0 and ADMM and set the standard for future studies which attempted to develop the next generation of blockchain which allowed not only for decentralized financial and information transactions, but also for autonomous operation of power systems.

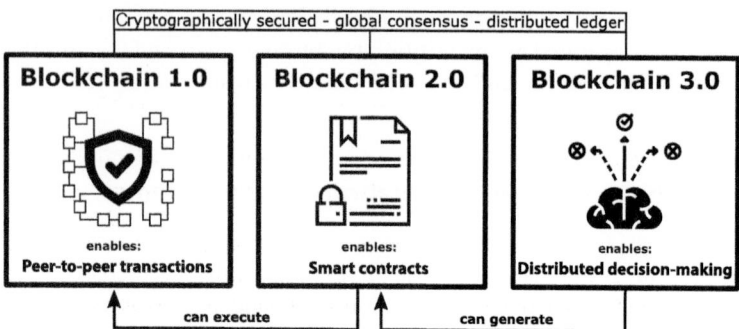

FIG. 4 Timeline of major milestones and milestones taking place in the transition toward blockchain-based energy trading. *(Icons licensed as CCBY: Blockchain by Maria Kislitsina, smart contract by Anatolii Babii, and decision making by Chanut is Industries from the Noun Project.)*

This set the stage for the development of what came to be known as Blockchain 3.0 platforms. The evolution of blockchain (Fig. 4) can thus be summarized as follows:

- **Blockchain 1.0**: A fully distributed ledger of transactions which are cryptographically secured and rely on global consensus.
- **Blockchain 2.0**: Includes smart contracts which digitally written and signed awaiting satisfaction of certain conditions to come into effect, executing peer-to-peer transactions.
- **Blockchain 3.0**: A fully decentralized platform capable of autonomous operation relying on distributed mathematical models. This self-managing system can determine optimal strategies to ensure global benefit, and thereby constructing smart contracts accordingly.

In this book, different blockchain-based solutions for the management of SGs are presented, ranging from trading markets to complex operational problems such as DR implementation and grid control. Thereby, all the solutions presented in this book are the state of the art in blockchain-based energy systems, being Blockchain 3.0 solutions. The ultimate objective of employing Blockchain 3.0 is to achieve the ideal structure of an IoE transactive energy network, possessing fully autonomous and fully decentralized operation, aiming at the benefit of end users first and foremost.

Acknowledgments

M. Lotfi acknowledges the support of the MIT Portugal Program (in Sustainable Energy Systems) by Portuguese funds through FCT, under grant PD/BD/142810/2018. J.P.S. Catalão acknowledges the support by FEDER funds through COMPETE 2020 and by Portuguese funds through FCT, under POCI-01-0145-FEDER-029803 (02/SAICT/2017).

References

[1] P.S. Antón, R. Silberglitt, J. Schneider, The global technology revolution: bio/nano/materials trends and their synergies with information technology by 2015, Tech. Rep. National Defense Research Institute, 2015, http://www.rand.org/.

[2] M. Hermann, T. Pentek, B. Otto, Design principles for Industrie 4.0 scenarios, in: Proceedings of the Annual Hawaii International Conference on System Sciences, vol. 2016-March, 2016, pp. 3928–3937, https://doi.org/10.1109/HICSS.2016.488.

[3] K. Schwab, The fourth industrial revolution: what it means and how to respond, World Economic Forum (2016), https://doi.org/10.1038/nnano.2015.286.

[4] L. Elliott, Fourth industrial revolution brings promise and peril for humanity, The Guardian (2016), http://www.theguardian.com/business/economics-blog/2016/jan/24/4th-industrial-revolution-brings-promise-and-peril-for-humanity-technology-davos.

[5] International Telecommunication Union, The Internet of Things, Tech. Rep., International Telecommunication Union, 2005, https://doi.org/10.1109/IEEESTD.2007.373646.

[6] A. Nordrum, Popular Internet of Things forecast of 50 billion devices by 2020 is outdated, in: IEEE Spectrum, 2016.

[7] Gartner, Gartner says 6.4 billion connected "Things" will be in use in 2016, Up 30 percent from 2015 [press release], Tech. Rep., Gartner, Inc., Barcelona, 2015, http://www.gartner.com/newsroom/id/3165317.

[8] Ericsson, Ericsson mobility report: on the pulse of the networked society, Tech. Rep., Ericsson, 2016, https://doi.org/10.3103/S0005105510050031.

[9] S. Lucero, IoT platforms: enabling the Internet of Things, in: IHS Technology, 2016, https://cdn.ihs.com/www/pdf/enabling-IOT.pdf.

[10] C. MacGillivray, M. Torchia, M. Cinco, M. Kalal, M. Kumar, R. Membrila, A. Siviero, Y. Torisu, N. Wallis, Worldwide Internet of Things forecast update, 2016–2020, IDC Market Forecast (2017) 40755516, http://www.idc.com/getdoc.jsp?containerId=US40755516.

[11] Dimensional Research, ParStream, Internet of Things (IoT) meets big data and analytics: a survey of IoT stakeholders, Tech. Rep., 2015 (march).

[12] A. Meola, The roles of cloud computing and fog computing in the Internet of Things revolution, in: Business Insider, 2016.

[13] IBM, What is cloud computing?, 2017, https://www.ibm.com/cloud-computing/learn-more/what-is-cloud-computing/.

[14] Gartner, Gartner says worldwide IoT security spending to reach $348 million in 2016 [press release], Tech. Rep., Gartner, Inc., 2016, http://www.gartner.com/newsroom/id/3291817.

[15] European Parliament and of the Council of 27 April 2016, Regulation (EU) 2016/679, 2016, https://eur-lex.europa.eu/eli/reg/2016/679/oj.

[16] A. Jung, Smart Grid 2.0 Building the Internet of Energy Supply, Spiegel Online, 2010, pp. 1–3.

[17] A.S. Sani, D. Yuan, J. Jin, L. Gao, S. Yu, Z.Y. Dong, Cyber security framework for Internet of Things-based energy internet, Future Gener. Comput. Syst. (2018), https://doi.org/10.1016/j.future.2018.01.029.

[18] C.C. Lin, D.J. Deng, W.Y. Liu, L. Chen, Peak load shifting in the internet of energy with energy trading among end-users, IEEE Access 5 (2017) 1967–1976, https://doi.org/10.1109/ACCESS.2017.2668143.

[19] Q. Sun, R. Han, H. Zhang, J. Zhou, J.M. Guerrero, A multi-agent-based consensus algorithm for distributed coordinated control of distributed generators in the energy internet, IEEE Trans. Smart Grid 6 (6) (2015) 1–14, https://doi.org/10.1109/TSG.2015.2412779.

[20] N. Bui, A.P. Castellani, P. Casari, M. Zorzi, The internet of energy: a web-enabled smart grid system, IEEE Netw. 26 (4) (2012) 39–45, https://doi.org/10.1109/MNET.2012.6246751.

[21] A. Jindal, N. Kumar, M. Singh, A unified framework for big data acquisition, storage and analytics for demand response management in smart cities, Future Gener. Comput. Syst. (2018), https://doi.org/10.1016/j.future.2018.02.039.

[22] M. Jaradat, M. Jarrah, A. Bousselham, Y. Jararweh, M. Al-Ayyoub, The internet of energy: smart sensor networks and big data management for smart grid, Procedia Comput. Sci. 56 (1) (2015) 592–597, https://doi.org/10.1016/j.procs.2015.07.250.

[23] K. Wang, J. Yu, Y. Yu, Y. Qian, D. Zeng, S. Guo, Y. Xiang, J. Wu, A survey on energy internet: architecture, approach, and emerging technologies, IEEE Syst. J. (2017) 1–14, https://doi.org/10.1109/JSYST.2016.2639820.

[24] P. Tenti, T. Caldognetto, Optimal control of local area energy networks (E-LAN), Sustain. Energy Grids Netw. 14 (2018) 12–24, https://doi.org/10.1016/j.segan.2018.03.002.

[25] J. Cao, M. Yang, Energy internet—towards Smart Grid 2.0, in: Proceedings of the International Conference on Networking and Distributed Computing, ICNDC, 2014, pp. 105–110, https://doi.org/10.1109/ICNDC.2013.10.

[26] S. Nakamoto, Bitcoin: A Peer-to-Peer Electronic Cash System, 2008, https://doi.org/10.1007/s10838-008-9062-0.

[27] M. Swan, Blockchain: Blueprint for a New Economy, O'Reilly Media, Inc., 2015, https://doi.org/10.1017/CBO9781107415324.004

[28] European Central Bank, Virtual currency schemes, Tech. Rep., 2012, http://www.ecb.europa.eu.

[29] European Central Bank, Virtual currency schemes—a further analysis, Tech. Rep., European Central Bank, 2015, https://doi.org/10.2866/662172.

[30] M. Lotfi, C. Monteiro, M. Shafie-Khah, J.P.S. Catalao, Evolution of demand response: a historical analysis of legislation and research trends, in: 2018 Twentieth International Middle East Power Systems Conference (MEPCON), December, IEEE, 2018, pp. 968–973, https://doi.org/10.1109/MEPCON.2018.8635264.

[31] Electric Power Research Institute, The Integrated Grid: Realizing the Full Value of Central and Distributed Energy Resources, EPRI, 2014.

[32] W. Cox, T. Considine, Structured energy: microgrids and autonomous transactive operation, in: 2013 IEEE PES Innovative Smart Grid Technologies Conference (ISGT), February, IEEE, 2013, pp. 1–6, https://doi.org/10.1109/ISGT.2013.6497919.

[33] W. Inam, D. Strawser, K.K. Afridi, R.J. Ram, D.J. Perreault, Architecture and system analysis of microgrids with peer-to-peer electricity sharing to create a marketplace which enables energy access, in: 2015 9th International Conference on Power Electronics and ECCE Asia (ICPE-ECCE Asia), June, 2015, pp. 464–469, https://doi.org/10.1109/ICPE.2015.7167826.

[34] M. Pustišek, A. Kos, U. Sedlar, Blockchain based autonomous selection of electric vehicle charging station, in: 2016 International Conference on Identification, Information and Knowledge in the Internet of Things (IIKI), October, 2016, pp. 217–222, https://doi.org/10.1109/IIKI.2016.60.

[35] N.Z. Aitzhan, D. Svetinovic, Security and privacy in decentralized energy trading through multi-signatures, blockchain and anonymous messaging streams, IEEE Trans. Depend. Secure Comput. 15 (5) (2018) 840–852, https://doi.org/10.1109/TDSC.2016.2616861.

[36] PwC, Blockchain—an opportunity for energy producers and consumers? Tech. Rep., PwC, 2016, http://www.pwc.com/gx/en/industries/energy-utilities-mining/power-utilities/publications/opportunity-for-energy-producers.html.

[37] German Energy Agency, Blockchain in the energy transition. A survey among decision-makers in the German energy industry, Tech. Rep., German Energy Agency, Berlin, 2016.

[38] J. Mattila, T. Seppälä, C. Naucler, R. Stahl, M. Tikkanen, A. Bådenlid, J. Seppälä, Industrial blockchain platforms: an exercise in use case development in the energy industry, Tech. Rep., The Research Institute of the Finnish Economy (ETLA), 2016, https://doi.org/10.1017/CBO978 1107415324.004.

[39] J. Horta, D. Kofman, D. Menga, Novel paradigms for advanced distribution grid energy management, ArXiv, 2017, abs/1712.05841.

[40] J. Green, P. Newman, Citizen utilities: the emerging power paradigm, Energy Policy 105 (2017) 283–293, https://doi.org/10.1016/j.enpol.2017.02.004.

[41] E. Munsing, J. Mather, S. Moura, Blockchains for decentralized optimization of energy resources in microgrid networks, in: 2017 IEEE Conference on Control Technology and Applications (CCTA), 2017, pp. 2164–2171, https://doi.org/10.1109/CCTA.2017.8062773.

Chapter 4

Decentralized operation of interdependent power and energy networks: Blockchain and security

M. Hadi Amini

School of Computing and Information Sciences, Florida International University, Miami, FL, United States; Sustainability, Optimization, and Learning for Interdependent Networks Laboratory (Solid Lab), Florida International University, Miami, FL, United States

1. Introduction

Internet of things (IoT) is increasingly emerging by integration of smart devices and communication technologies. Cisco Inc. predicted that by 2020 the number of connected devices will be about 50 billion predicting to have 50 billion [1]. In this context, blockchain is introduced as an effective means of decentralizing transactions. Blockchain is deployed to enable peer-to-peer transactions using "distributed ledger technology" [2]. Recent studies emphasized on the pivotal role of smart contracts in secure and decentralized energy trading in smart grids [3–12]. IoT covers a wide range of networks, including smart city infrastructures, such as interdependent power and energy networks [13]. Integration of smart grid technologies further leads to more distributed resources [14–16], such as microgrids [17, 18], distributed generation units [19], smart meters [20, 21], and electric vehicles (EVs) [22–24]. In this chapter, I will first provide a detailed insight on the transition from centralized energy systems toward decentralized operating paradigm. This change of paradigm requires enhanced infrastructures and algorithms to deal with the large-scale nature of the power systems as well as ensuring efficient operation of the system. I further explain how decentralized transactions among heterogeneous agents in smart power grids, enabled by blockchain, can contribute to this decentralized operation of power systems in a more secure and efficient way.

From the methodological viewpoint, there is a transition from centralized decision-making toward distributed agent-based optimization. In both application

Blockchain-based Smart Grids. https://doi.org/10.1016/B978-0-12-817862-1.00004-X

domains which are being investigated in this thesis (transportation electrification and state estimation), we address the planning stage in a centralized fashion. Although centralized solutions are effective for the planning stage, when there is sufficient time to solve the large-scale optimization problems, there is a crucial need for fast optimization algorithms to enable real-time decision-making. This is the key motivation for deploying distributed optimization techniques to enable optimal operation of emerging systems, that is, distributed charge coordination of EVs considering power system constraints, and distributed state estimation for situational awareness in power distribution systems.

2. Grid modernization toward smarter power systems

Conventionally, power systems operate in a unidirectional manner, that is, power flows from the generation side to the end users through transmission and distribution systems. Further, as end users only act as power system consumers, they do not need to have access to the generation information and measurement data, that is, most of sensing devices are installed at the distribution side, collect information, and share the information with operators for optimal energy dispatch. Fig. 1 represents this structure, as well as information and power flow.

There is a paradigm change toward bidirectional information and power flow. Growing integration of modern technologies, such as EVs [25], distribution automation [26], intelligent sectionalizers, storage [27], distributed generation [28], microgrids [29, 30], smart meters, and renewable energy resources [31] has recently seen accelerated adoption in smart power distribution networks (PDNs). These technologies enable power injection at the distribution systems and end-user level. Further, emerging demand-side management strategies and demand response programs enable participation of end users in

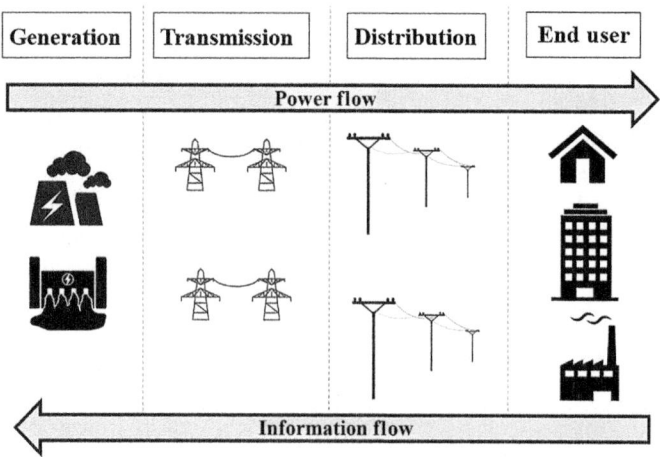

FIG. 1 Unidirectional information and power flow in conventional power system.

balancing load and generation [32–34]. In order to make a decision in terms of optimal participation level in the energy market, end users need to have access to more information as compared with the conventional power systems. Although demand-side resources can contribute to the operation of power networks, there should be proper cost-benefit analysis to ensure feasibility and cost effectiveness of this participation. For instance, vehicle-to-grid (V2G) technologies are introduced as promising technologies to incentivize EV drivers toward contributing to grid storage. However, according to [35, 36], there is not likely sufficient incentives for EV drivers to enable large-scale participation of EV batteries as energy storage units. Further, emerging (ultra)fast charging technologies may have destructive effects on PDNs due to increasing peak load demand [37], high-power density, and stochastic nature of charging demand [38]. Given the mentioned issues introduced by fast charging stations [37] and feasibility of V2G implementation [35], in this thesis, we focus on unidirectional charge scheduling, that is, EVs are only acting as flexible loads rather than energy storage units. Further, in order to address the concerns regarding the higher peak load demand caused by fast charging rates, we include two additional constraints in our charge scheduling optimization problem to model charging rate limit of each EV and power limit of each charging station. Fig. 2 represents the smart power system paradigm and corresponding information and power flow. As this schematic overview shows, both power and information can flow in both directions among the generation side and end users. On one hand, this transition and paradigm change can enhance the operation of power systems by achieving redundancy employing more distributed energy resources. On the other hand, increasing the number of distributed energy resources increases the computational complexity of the optimization problems in the context of PDN operation.

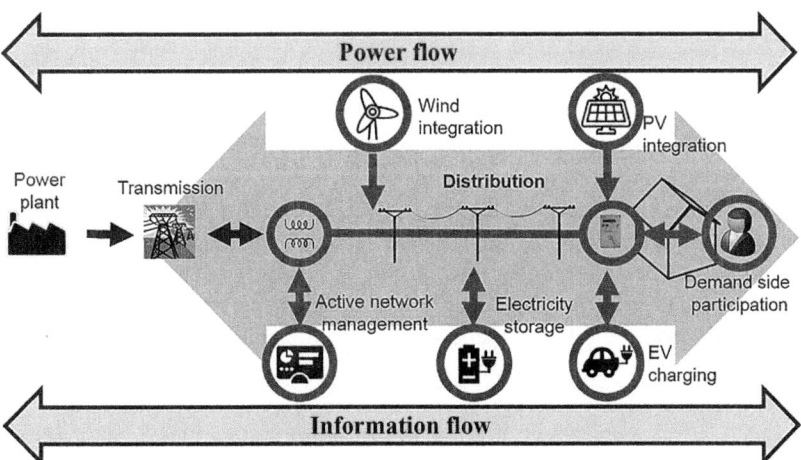

FIG. 2 Bidirectional information and power flow in smart power system. *(Courtesy of http://www. edsoforsmartgrids.eu/home/why-smart-grids/.)*

Further, ensuring the cybersecurity of PDNs is a more challenging task due to distributed nature of emerging resources.

Although previous studies focused on integrating smart grid technologies independently, there is a need to present an inclusive integration model that is capable of managing these technologies simultaneously as well as taking into account physical network constraints from the distribution network utility point of view. In practice, as these resources need to operate at the same time, we need a more integrated approach that takes advantage of their flexibility to enable the proper integration and improve the operation of the network. For instance, plug-in electric vehicles (PEVs) can introduce spatiotemporal flexibility in terms of charge scheduling, if properly integrated. They have the potential to enable much needed flexibility and dispatchability to the power network. Coordination of these resources, including PEVs, while maintaining the optimal operation of the physical network (e.g., congestion management and feasible power flow solution) is not addressed effectively. Similarly, current solutions based on mostly centralized decision making may not be scalable for the large-scale PDNs. Previous studies, in this context, mostly suffer from two main drawbacks: (1) they mostly focus on independent integration of these resources [25, 29, 31] and (2) some studies neglect the physical network constraints while proposing the optimal coordination methods, such as optimal PEV management [39].

3. Large-scale problems: From centralized to decentralized optimization

Power systems deploy a large number of agents for their economic and reliable operation. According to [40], the US eastern interconnection operates based on about 100 control areas and approximately 50,000 buses. These areas and buses are managed by thousands of intelligent and mechanical agents [40]. This increases the complexity of power systems due to the computational complexity of power systems operation, making the central decision making for optimal operation an impossible task [40]. In order to solve large-scale optimization and situational awareness problems in power systems, prior studies have deployed various distributed techniques, including, distributed semidefinite programming [41], alternating direction method of multipliers [42], optimality condition decomposition [43], multiagent systems [14], Lagrangian relaxation [16], proximal message passing [44], auxiliary problem principle [45], diagonal quadratic approximation [46], distributed model predictive control [47], and consensus + innovations fully distributed optimization [48]. There are three major structures for optimization algorithms: centralized, decentralized/hierarchical, and distributed.

Fig. 3 represents the centralized optimization structure. In this structure, all agents collect the data through various means, such as distributed sensors and measurement devices. They will then share all information with a fusion center

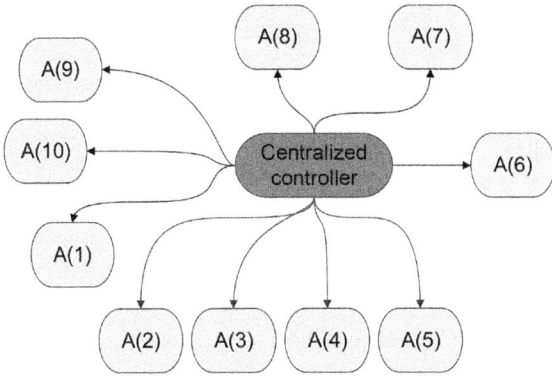

Centralized structure

FIG. 3 Overview of centralized optimization structure.

for central decision making. The fusion center solves a large-scale optimization problem considering the objective of the whole system. The results of this optimization problem are control signals for each agent. These control signals are broadcast to agents so that they can operate based on the optimal operation point of the entire system.

Fig. 4 illustrates decentralized/hierarchical optimization structure. Most of these methods in the literature require a fusion center in the so-called distributed structure. In this chapter, we distinguish decentralized and distributed methods by identifying the existence of fusion center. In the power system literature, some studies deployed decentralized algorithms for various problems, including optimal power flow [49–52], state estimation [53], and power system protection [54]. System of systems-based methods lend themselves as decentralized efficient methods for power system applications [55].

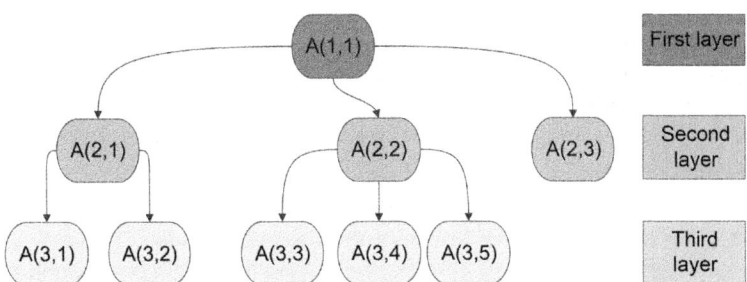

Hierarchical/decentralized structure

FIG. 4 Overview of decentralized/hierarchical optimization structure.

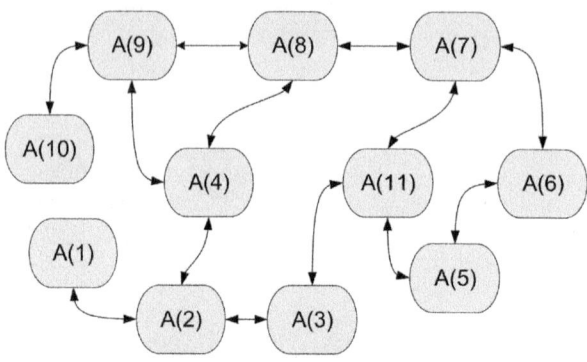

Distributed structure

FIG. 5 Overview of distributed optimization structure.

Fig. 5 represents a fully distributed optimization structure. Fully distributed methods do not need a fusion center. This enables fast plug-and-play capability for real-time decision-making purposes.

4. Blockchain-based interdependent power and energy network: A transition toward decentralized, smart, and secure energy trading

Here, I try to review the highlights of previous studies on blockchain-based smart grids. In [56], an agent-based simulation of blockchain protocols is demonstrated. To this end, they define Chainweb through a network of miners that keep a local version of their arboretum. Further, a simulation platform is developed to analyze the interaction of Byzantine and economic-aware adversaries with the protocols [56].

Blockchain is also considered as a distributed technology to enable financial transactions using a series of blocks, eliminating the need for a trusted third party [57]. In [57], several applications of blockchain are explored in several categories, including the IoT, finance, cybersecurity, smart property, health care, smart contracts, and wireless networks. In this context, cyberphysical systems (such as energy systems and vehicular systems) are categorized under the IoT. In this chapter, we focus on application of blockchain in decentralized operation of interdependent power and energy networks. Vehicle-to-grid (V2G) technologies are evolving to ensure bidirectional power flow between EVs and power grid [58, 59]. In this context, deploying blockchain paves the road for energy trading among entities, while implementing V2G technologies. Zhou et al. [60] used edge computing and blockchain technologies to enable secure energy exchange for V2G. Their approach has two layers: (1) deploying blockchain for V2G energy exchange and (2) deploying edge computing to

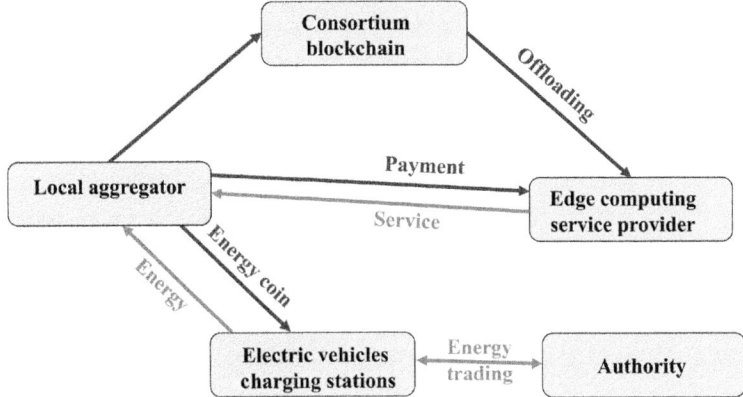

FIG. 6 Overview of the proposed approach in [60].

improve performance of local aggregator [60]. An overview of the proposed approach in [60] is illustrated in Fig. 6.

In [61], smart metering infrastructure is leveraged as secured entity for energy data exchange. Further, a blockchain-based approach is proposed for energy exchange among communities with solar energy generation [61]. An overview of the proposed approach in [61] is illustrated in Fig. 7.

Vishal [62] proposed an efficient approach to optimize the energy consumption of blockchain-driven Internet of vehicles. To this end, the number of information exchange requirements, in terms of transactions, is optimized using distributed clustering [62]. Further, Pee et al. develop peer-to-peer systems

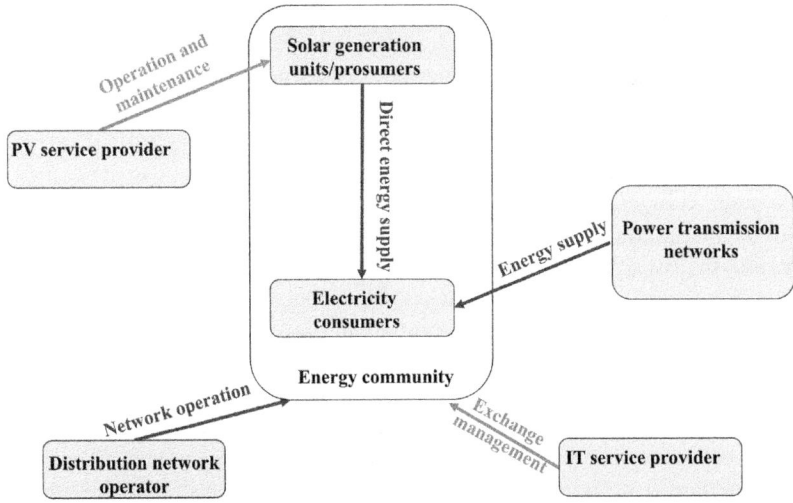

FIG. 7 Overview of the proposed approach in [61].

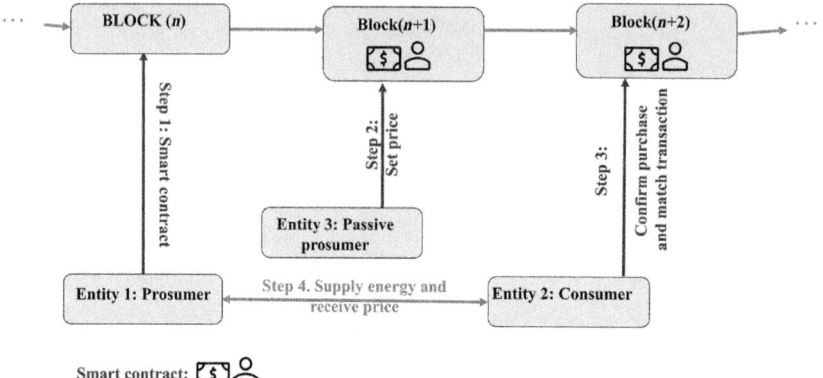

FIG. 8 Overview of the proposed approach in [64].

for distributed energy trading [63]. They assigned one block to each prosumer in the smart energy network to enable peer-to-peer contracts, leading to distributed energy market [63]. They developed another framework for blockchain-based energy trading via smart contracts considering renewable resources [64]. They used Ethereum [65] for the smart contracts. An overview of the proposed approach in [64] is illustrated in Fig. 8.

In [66], energy blockchain is deployed for secure operation of EVs in smart community. To this end, execution of smart contracts for EV charging is enabled using verified energy blockchain. This study used delegated Byzantine fault tolerance consensus algorithm to reach an agreement between blockchain agents, while analyzing optimal contracts using contract theory to meet the energy demand of all EVs and optimizing the utility of the operator [66]. In [67], a permissioned blockchain is used to enable secure and decentralized energy trading among renewable energy resources. According to this study, blockchain introduces several advantages by enabling verifiablity of transactions, distributedness, security, and reducing third parties [67]. Chen et al. [68] define master-slave structure for blockchain to ensure security of power systems. In [69], a blockchain-based algorithm is proposed to incentivize EVs in the vehicular energy network toward specific goals. Blockchain does not only enable distributed implementation of this study but also ensures security [69]. In [70], a novel direction is followed in terms of using blockchain for smart microgrids. Authors used blockchain not only to enable distributed economic decisions but also to enable distributed operation of microgrids. An overview of the transition from centralized to decentralized, and ultimately to distributed peer-to-peer structure based on [71] is illustrated in Fig. 9.

Pipattanasomporn et al. [72] deployed blockchain using Hyperledger to enable solar energy exchange among consumers. According to [73], depending on computing power of agents who are participating in distributed blockchain-based transactions, they may be able to manipulate the data of transaction, stop transaction verification, or prevent mining at one of the current blocks.

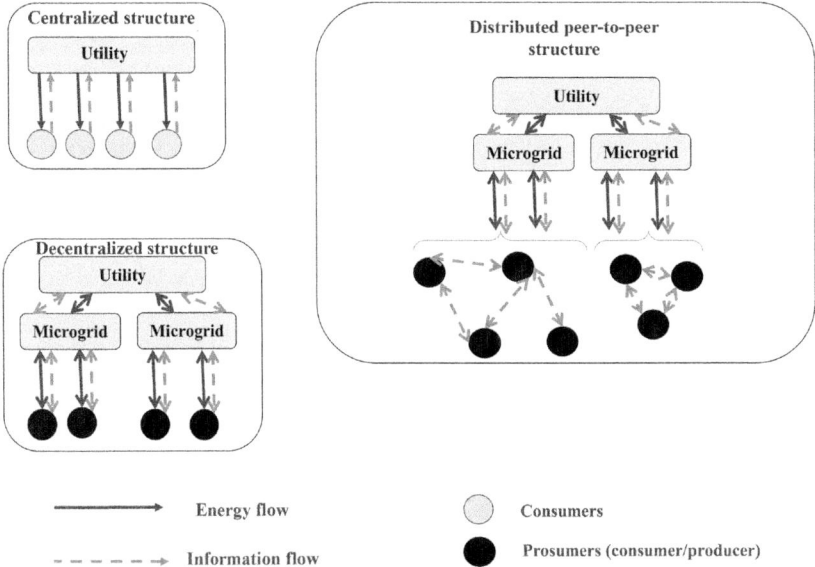

FIG. 9 Overview of the transition from centralized structure to peer-to-peer structure in [71].

5. Conclusion

In this chapter, I first explained the transition from central operation of critical smart city infrastructures toward decentralized resources and IoT agents, such as distributed energy resources, microgrids, and smart mobility sensing devices. I further investigated the power grid modernization toward smarter power systems. Smart grids introduce more reliable, secure, efficient, and sustainable energy delivery to the consumers by engaging demand side in load-generation balance. Finally, I introduced blockchain-based energy trading as an effective means of boosting the decentralization of interdependent power and energy networks by enabling decentralized transactions and smart contracts. To this end, I provided an overview of the prior work on blockchain-based smart grids, as well as different frameworks for implementing decentralized energy exchange among power and energy system agents in a decentralized and secure fashion.

References

[1] M.A. Khan, K. Salah, IoT security: review, blockchain solutions, and open challenges, Future Gener. Comput. Syst. 82 (2018) 395–411.

[2] P. Siano, G. De Marco, A. Rolán, V. Loia, A survey and evaluation of the potentials of distributed ledger technology for peer-to-peer transactive energy exchanges in local energy markets, IEEE Syst. J. 13 (2019) 3454–3466.

[3] K. Gai, Permissioned blockchain and edge computing empowered privacy-preserving smart grid networks, IEEE Internet Things J. 6 (2019) 7992–8004.

[4] H. Zhang, J. Wang, Y. Ding, Blockchain-based decentralized and secure keyless signature scheme for smart grid, Energy 180 (2019) 955–967.

[5] K. Valtanen, J. Backman, S. Yrjölä, Blockchain-powered value creation in the 5G and smart grid use cases, IEEE Access 7 (2019) 25690–25707.

[6] J. Lin, M. Pipattanasomporn, S. Rahman, Comparative analysis of blockchain-based smart contracts for solar electricity exchanges, in: 2019 IEEE Power & Energy Society Innovative Smart Grid Technologies Conference, 2019.

[7] R. Chaudhary, BEST: Blockchain-based secure energy trading in SDN-enabled intelligent transportation system, Comput. Secur. 85 (2019) 288–299.

[8] M.R. Alam, M. St-Hilaire, T. Kunz, Peer-to-peer energy trading among smart homes, Appl. Energy 238 (2019) 1434–1443.

[9] W. Tushar, A motivational game-theoretic approach for peer-to-peer energy trading in the smart grid, Appl. Energy 243 (2019) 10–20.

[10] Z. Guan, Privacy-preserving and efficient aggregation based on blockchain for power grid communications in smart communities, IEEE Commun. Mag. 56.7 (2018) 82–88.

[11] S. Noor, Energy demand side management within micro-grid networks enhanced by blockchain, Appl. Energy 228 (2018) 1385–1398.

[12] Z. Su, Y. Wang, Q. Xu, M. Fei, Y.C. Tian, N. Zhang, A secure charging scheme for electric vehicles with smart communities in energy blockchain, IEEE Internet Things J. (2018).

[13] M.H. Amini, H. Arasteh, P. Siano, Sustainable smart cities through the lens of complex interdependent infrastructures: panorama and state-of-the-art, in: M. Amini, K. Boroojeni, S. Iyengar, P. Pardalos, F. Blaabjerg, A. Madni (Eds.), Sustainable Interdependent Networks II, Springer, Cham, 2019, pp. 45–68.

[14] M.H. Amini, B. Nabi, M.-R. Haghifam, Load management using multi-agent systems in smart distribution network, in: 2013 IEEE Power & Energy Society General Meeting, IEEE, 2013, pp. 1–5.

[15] S. Bahrami, M.H. Amini, A decentralized trading algorithm for an electricity market with generation uncertainty, Appl. Energy 218 (2018) 520–532.

[16] M.H. Amini, S. Bahrami, F. Kamyab, S. Mishra, R. Jaddivada, K. Boroojeni, P. Weng, Y. Xu, Decomposition methods for distributed optimal power flow: panorama and case studies of the DC model, in: A.F. Zobaa, S.H.E. Abdel Aleem, A.Y. Abdelaziz (Eds.), Classical and Recent Aspects of Power System Optimization, Elsevier, 2018, pp. 137–155.

[17] M.H. Amini, K.G. Boroojeni, T. Dragičević, A. Nejadpak, S.S. Iyengar, F. Blaabjerg, A comprehensive cloud-based real-time simulation framework for oblivious power routing in clusters of DC microgrids, in: 2017 IEEE Second International Conference on DC Microgrids (ICDCM), 2017, pp. 270–273.

[18] A. Mohammadi, S. Bahrami, An overview of future microgrids, in: S. Bahrami, A. Mohammadi (Eds.), Smart Microgrids,Springer, Cham, 2019, pp. 1–6.

[19] A.R. Al-Ali, R. Aburukba, Role of Internet of things in the smart grid technology, J. Comput. Commun. 3 (5) (2015) 229.

[20] J. Lloret, An integrated IoT architecture for smart metering, IEEE Commun. Mag. 54 (12) (2016) 50–57.

[21] R.W.R. de Souza, Deploying wireless sensor networks-based smart grid for smart meters monitoring and control, Int. J. Commun. Syst. 31 (10) (2018) e3557.

[22] M.H. Amini, O. Karabasoglu, Optimal operation of interdependent power systems and electrified transportation networks, Energies 11 (1) (2018) 196.

[23] M.H. Amini, M.P. Moghaddam, O. Karabasoglu, Simultaneous allocation of electric vehicles' parking lots and distributed renewable resources in smart power distribution networks, Sustain. Cities Soc. 28 (2017) 332–342.

[24] M.H. Amini, A. Kargarian, O. Karabasoglu, ARIMA-based decoupled time series forecasting of electric vehicle charging demand for stochastic power system operation, Electr. Pow. Syst. Res. 140 (2016) 378–390.

[25] J.A.P. Lopes, F.J. Soares, P.M.R. Almeida, Integration of electric vehicles in the electric power system, Proc. IEEE 99 (1) (2011) 168–183.

[26] R.E. Brown, Impact of smart grid on distribution system design, in: 2008 IEEE Power and Energy Society General Meeting-Conversion and Delivery of Electrical Energy in the 21st Century, IEEE, 2008, pp. 1–4.

[27] N.S. Wade, P. Taylor, P. Lang, P. Jones, Evaluating the benefits of an electrical energy storage system in a future smart grid, Energy Policy 38 (11) (2010) 7180–7188.

[28] H. Farhangi, The path of the smart grid, IEEE Power Energy Mag. 8 (1) (2010) 18–28.

[29] S. Chowdhury, P. Crossley, Microgrids and Active Distribution Networks, The Institution of Engineering and Technology, 2009.

[30] J. Zhang, M.H. Amini, P. Weng, A hierarchical approach based on the Frank-Wolfe algorithm and Dantzig-Wolfe decomposition for solving large economic dispatch problems in smart grids, in: S. Bahrami, A. Mohammadi (Eds.), Smart Microgrids, Springer, 2019, pp. 41–56.

[31] A.M.L. da Silva, L.C. Nascimento, M.A. da Rosa, D. Issicaba, J.A.P. Lopes, Distributed energy resources impact on distribution system reliability under load transfer restrictions, IEEE Trans. Smart Grid 3 (4) (2012) 2048–2055.

[32] F. Kamyab, M.H. Amini, S. Sheykhha, M. Hasanpour, M.M. Jalali, Demand response program in smart grid using supply function bidding mechanism, IEEE Trans. Smart Grid 7 (3) (2015) 1277–1284.

[33] M.H. Amini, S. Talari, H. Arasteh, N. Mahmoudi, M. Kazemi, A. Abdollahi, V. Bhattacharjee, M. Shafie-Khah, P. Siano, J.P. Catalão, Demand response in future power networks: panorama and state-of-the-art, in: M. Amini, K. Boroojeni, S. Iyengar, P. Pardalos, F. Blaabjerg, A. Madni (Eds.), Sustainable Interdependent Networks II, Springer, 2019, pp. 167–191.

[34] A. Imteaj, M.H. Amini, J. Mohammadi, Leveraging decentralized artificial intelligence to enhance resilience of energy networks, arXiv preprint arXiv:1911.07690 (2019).

[35] S.B. Peterson, J. Apt, J. Whitacre, Lithium-ion battery cell degradation resulting from realistic vehicle and vehicle-to-grid utilization, J. Power Sources 195 (8) (2010) 2385–2392.

[36] S.B. Peterson, J. Whitacre, J. Apt, The economics of using plug-in hybrid electric vehicle battery packs for grid storage, J. Power Sources 195 (8) (2010) 2377–2384.

[37] Y.H. Febriwijaya, A. Purwadi, A. Rizqiawan, N. Heryana, A study on the impacts of DC fast charging stations on power distribution system, in: 2014 International Conference on Electrical Engineering and Computer Science (ICEECS), IEEE, 2014, pp. 136–140.

[38] D. Meyer, J. Wang, Integrating ultra-fast charging stations within the power grids of smart cities: a review, IET Smart Grid 1 (1) (2018) 3–10.

[39] Y. Cao, S. Tang, C. Li, P. Zhang, Y. Tan, Z. Zhang, J. Li, An optimized EV charging model considering TOU price and SOC curve, IEEE Trans. Smart Grid 3 (1) (2012) 388–393.

[40] P. Hines, S. Talukdar, Controlling cascading failures with cooperative autonomous agents, Int. J. Crit. Infrastruct. 3 (1) (2007).

[41] H. Zhu, G.B. Giannakis, Power system nonlinear state estimation using distributed semidefinite programming, IEEE J. Sel. Top. Sign. Process. 8 (6) (2014) 1039–1050.

[42] T. Erseghe, Distributed optimal power flow using ADMM, IEEE Trans. Power Syst. 29 (5) (2014) 2370–2380.

[43] A. Rabiee, B. Mohammadi-Ivatloo, M. Moradi-Dalvand, Fast dynamic economic power dispatch problems solution via optimality condition decomposition, IEEE Trans. Power Syst. 29 (2) (2014) 982–983.

[44] M. Kraning, E. Chu, J. Lavaei, S. Boyd, Dynamic Network Energy Management Via Proximal Message Passing, vol. 1, Foundations and Trends(R), 2014, pp. 73–126.

[45] D. Hur, J. Park, B.H. Kim, Evaluation of convergence rate in the auxiliary problem principle for distributed optimal power flow, in: IEE Proceedings—Generation, Transmission, and Distribution, vol. 149, IET, 2002, pp. 525–532.

[46] A. Mohammadi, M. Mehrtash, A. Kargarian, Diagonal quadratic approximation for decentralized collaborative TSO+DSO optimal power flow, IEEE Trans. Smart Grid 10 (3) (2018) 2358–2370.

[47] E. Camponogara, D. Jia, B.H. Krogh, S. Talukdar, Distributed model predictive control, IEEE Control Syst. Mag. 22 (1) (2002) 44–52.

[48] M.H. Amini, J. Mohammadi, S. Kar, Distributed holistic framework for smart city infrastructures: tale of interdependent electrified transportation network and power grid, IEEE Access 7 (2019) 157535–157554.

[49] A.J. Conejo, J.A. Aguado, Multi-area coordinated decentralized DC optimal power flow, IEEE Trans. Power Syst. 13 (4) (1998) 1272–1278.

[50] G. Hug-Glanzmann, G. Andersson, Decentralized optimal power flow control for overlapping areas in power systems, IEEE Trans. Power Syst. 24 (1) (2009) 327–336.

[51] P.N. Biskas, A decentralized implementation of DC optimal power flow on a network of computers, IEEE Trans. Power Syst. 20 (1) (2005) 25–33.

[52] J. Shah, B.F. Wollenberg, N. Mohan, Decentralized power flow control for a smart micro-grid, in: IEEE Power and Energy Society General Meeting, 2011.

[53] V. Kekatos, G.B. Giannakis, Distributed robust power system state estimation, IEEE Trans. Power Syst. 28 (2) (2012) 1617–1626.

[54] N. Higgins, V. Vyatkin, N.K.C. Nair, K. Schwarz, Distributed power system automation with IEC 61850, IEC 61499, and intelligent control, IEEE Trans. Syst. Man Cybern. Part C Appl. Rev. 41 (1) (2010) 81–92.

[55] A. Mohammadi, F. Safdarian, M. Mehrtash, A. Kargarian, A system of systems engineering framework for modern power system, in: Sustainable Interdependent Networks II. From Smart Power Grids to Intelligent Transportation Networks, 2019, p. 217.

[56] T. Chitra, M. Quaintance, S. Haber, W. Martino, Agent-based simulations of blockchain protocols illustrated via Kadena's Chainweb, in: 2019 IEEE European Symposium on Security and Privacy Workshops (EuroS&PW), 2019, pp. 386–395, Stockholm, Sweden.

[57] D.B. Rawat, V. Chaudhary, R. Doku, Blockchain: emerging applications and use cases, arXiv preprint arXiv:1904.12247 (2019).

[58] M.R. Mozafar, M.H. Amini, M.H. Moradi, Innovative appraisement of smart grid operation considering large-scale integration of electric vehicles enabling V2G and G2V systems, Electr. Power Syst. Res. 154 (2018) 245–256.

[59] M.H. Amini, A panorama of interdependent power systems and electrified transportation networks, in: Sustainable Interdependent Networks II, Springer, Cham, 2019, pp. 23–41.

[60] Z. Zhou, T. Lu, G. Xu, Blockchain and edge computing based vehicle-to-grid energy trading in energy internet, in: 2nd IEEE Conference on Energy Internet and Energy System Integration (EI2), IEEE, 2018.

[61] C. Plaza, Distributed solar self-consumption and blockchain solar energy exchanges on the public grid within an energy community, in: 2018 IEEE International Conference on Environment and Electrical Engineering and 2018 IEEE Industrial and Commercial Power Systems Europe (EEEIC/I&CPS Europe), 2018.

[62] V. Sharma, An energy-efficient transaction model for the blockchain-enabled Internet of vehicles (IoV), IEEE Commun. Lett. 23 (2) (2019) 246–249.

[63] S.J. Pee, E.S. Kang, J.G. Song, J.W. Jang, Blockchain based smart energy trading platform using smart contract, in: 2019 International Conference on Artificial Intelligence in Information and Communication (ICAIIC), 2019, pp. 322–325.

[64] E.S. Kang, S.J. Pee, J.G. Song, J.W. Jang, A blockchain-based energy trading platform for smart homes in a microgrid, in: 3rd International Conference on Computer and Communication Systems (ICCCS), IEEE, 2018, pp. 472–476.

[65] V. Buterin, Ethereum white paper: a next generation smart contract & decentralized application platform, 2014, first version.

[66] Y. Wang, Z. Su, Q. Xu, N. Zhang, Contract based energy blockchain for secure electric vehicles charging in smart community, in: 2018 IEEE 16th International Conference on Dependable, Autonomic and Secure Computing, 16th International Conference on Pervasive Intelligence and Computing, 4th International Conference on Big Data Intelligence and Computing and Cyber Science and Technology Congress (DASC/PiCom/DataCom/CyberSciTech), 2018, pp. 323–327.

[67] R.K. Kodali, S. Yerroju, B.Y.K. Yogi, Blockchain based energy trading, in: TENCON 2018–2018 IEEE Region 10 Conference, 2018, pp. 1778–1783.

[68] X. Chen, X. Hu, Y. Li, X. Gao, D. Li, A blockchain based access authentication scheme of energy internet, in: 2nd IEEE Conference on Energy Internet and Energy System Integration (EI2), 2018, pp. 1–9.

[69] Y. Wang, Z. Su, N. Zhang, BSIS: blockchain based secure incentive scheme for energy delivery in vehicular energy network, IEEE Trans. Ind. Inf. 15 (2019) 3620–3631.

[70] M.L. Di Silvestre, P. Gallo, M.G. Ippolito, E.R. Sanseverino, G. Zizzo, A technical approach to the energy blockchain in microgrids, IEEE Trans. Ind. Inf. 14 (11) (2018) 4792–4803.

[71] A. Ahl, M. Yarime, K. Tanaka, D. Sagawa, Review of blockchain-based distributed energy: implications for institutional development, Renew. Sustain. Energy Rev. 107 (2019) 200–211.

[72] M. Pipattanasomporn, M. Kuzlu, S. Rahman, A blockchain-based platform for exchange of solar energy: laboratory-scale implementation, in: 2018 International Conference and Utility Exhibition on Green Energy for Sustainable Development (ICUE), IEEE, 2018.

[73] I.-C. Lin, T.-C. Liao, A survey of blockchain security issues and challenges, Int. J. Netw. Secur. 19 (5) (2017) 653–659.

Further reading

W.F. Tinney, C.E. Hart, Power flow solution by Newton's method, IEEE Trans. Power App. Syst. 11 (1967) 1449–1460.

H.W. Dommel, W.F. Tinney, Optimal power flow solutions, IEEE Trans. Power App. Syst. 10 (1968) 1866–1876.

M.A. Abido, Optimal power flow using particle swarm optimization, Int. J. Electr. Power Energy Syst. 24 (7) (2002) 563–571.

F.C. Schweppe, J. Wildes, Power system static-state estimation, Part I: exact model, IEEE Trans. Power App. Syst. 1 (1970) 120–125.

A. Monticelli, Electric power system state estimation, Proc. IEEE 88 (2) (2000) 262–282.

C.W. So, K.K. Li, Time coordination method for power system protection by evolutionary algorithm, IEEE Trans. Ind. Appl. 36 (5) (2000) 1235–1240.

J. Bertsch, C. Carnal, D. Karlson, J. Mcdaniel, K. Vu, Wide-area protection and power system utilization, Proc. IEEE 93 (5) (2005) 997–1003.

Chapter 5

The role of various market participants in blockchain business model

Saber Talari[a], Hosna Khajeh[b], Miadreza Shafie-khah[b], Barry Hayes[c], Hannu Laaksonen[b] and João P.S. Catalão[d]

[a]*Fraunhofer Institute for Energy Economics and Energy System Technology, Kassel, Germany,* [b]*School of Technology and Innovations, University of Vaasa, Vaasa, Finland,* [c]*School of Engineering, University College Cork, Cork, Ireland,* [d]*Faculty of Engineering of the University of Porto and INESC TEC, Porto, Portugal*

1. Introduction

Decentralization of energy networks will bring generation units close to customers. This enables reduction of energy losses and emissions thanks to renewables, like photovoltaic systems (PVs) and wind turbine-based generation. Moreover, active utilization of distributed energy resources (DER) control potential can enable improvement of power system reliability and resiliency at both local (e.g., distribution network) and system-wide (e.g., transmission network) levels. Active and intelligent utilization of DER for local and system-wide needs is dependent on available enabling technologies and market and regulation schemes. For example, some countries have made some regulations to run peer-to-peer (P2P) electricity markets.

In addition to other applications, blockchain technology can be used to realize P2P markets in which transactions are done in a decentralized way without any requirements for a central entity like distribution system operator (DSO). Blockchain technology-based decentralized P2P market is able to handle hundreds or thousands of transactions almost in real time. For example, the blockchain algorithm of Bitcoin is able to handle seven transactions per second [1]. Therefore some other algorithms such as Ethereum with ability of handling tens of transactions per second or Hyperledger to deal with hundreds of energy transaction [2] can be employed, instead.

Some recent pilot projects have applied the aforementioned blockchain algorithms. For instance, a blockchain-based P2P energy trading in Brooklyn,

Blockchain-based Smart Grids. https://doi.org/10.1016/B978-0-12-817862-1.00005-1

75

USA, has been performed experimentally to buy and sell energy among prosumers by Ethereum platform for the smart contracts [3].

Accordingly, blockchain technology has potential to be able to integrate different market players in new and better way than today to provide, e.g., balancing services. Moreover, it can incentivized new players through special smart contracts. For example, electric vehicles (EVs) can be paid for their active role as energy supplier or demand response provider. As Fig. 5.1 shows, market participants in blockchain-based platform are communicating together in almost real time, and the platform is able to analyze big data that all leads to better trading system. Within this blockchain-based trading platform, participants can automatically settle contracts physically and financially.

For example, provision of energy balancing technical flexibility services for local (DSO) or system-wide (transmission system operators [TSO]) needs by different available flexible resources like EVs, demand response, energy storages, or distributed generation can be done either directly or through ancillary flexibility service markets as shown in Fig. 5.1. In traditional markets, ancillary service provision is done through hierarchical trading and market scheme in such a way where TSO confirms the compatibility of the service providers (i.e., flexibility sources) to participate in different ancillary service markets. Then, flexibility offers are sent to the TSO and related flexibility sources from different markets. However, in decentralized blockchain-based scheme, all

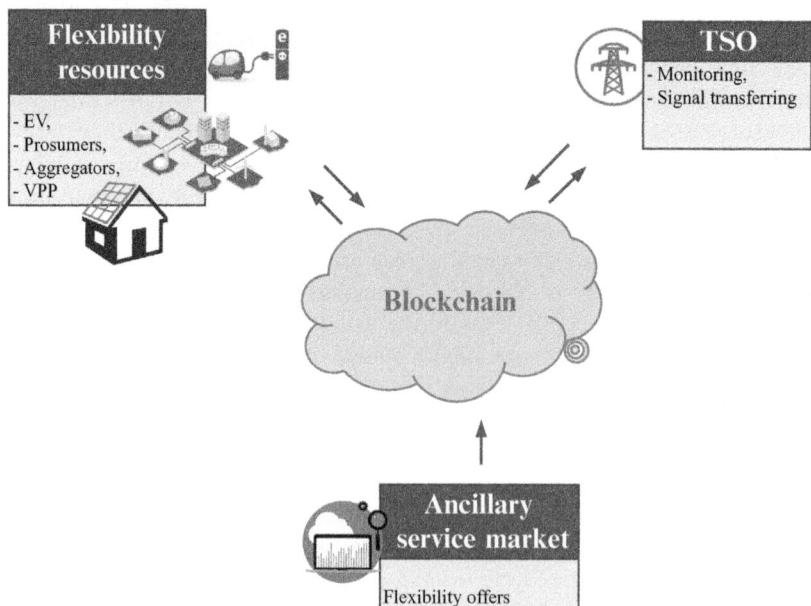

FIG. 5.1 Blockchain-based platform for flexibility trading with market participants.

participants are connected to each other directly. Therefore a transaction can be settled automatically without need of central entity or intermediary.

The rest of this chapter contains the following section. In Section 2, different market participants are introduced, while their role in the blockchain-based market platform is discussed. In Section 3, different possible business models for blockchain platform are presented, and the position of each market participants within the models is described. Finally, some remarks and brief results regarding this study are indicated in Section 4.

2. Market participants

2.1 Prosumers

An increasing number of customers can act in the future as a prosumer, i.e., be both producer and consumer. In fact, they are able to produce their own energy while consuming. Depending on the available management systems and storage solutions, prosumers can be able to produce their own energy that they need for their consumption. On the other hand, prosumer can be also defined as a household with DERs for self-consumption and extra production (generation). Depending on the available technologies and regulation, household prosumers can participate in different type of markets with different business models like P2P, flexibility trading, over-the-counter (OTC) trading, and crowdsale trading [4, 5].

In the user-centric control and market approaches, suitable incentives to achieve maximum simultaneous collaborative benefit for all different parties should be found and aimed at. From prosumers' viewpoint, it is important to have enough possibilities to offer their flexibility services to maximize own benefits. Therefore prosumers should be able to participate in local market or make a contract with an aggregator or flexible operator. Prosumers can have several flexibilities such as renewable energy resources (RES), controllable loads, battery storage, and EVs that enable them to be active players in the market [6]. If they would like to participate in a local market, retailer price can be upper bound for them to pay, and the feed-in tariffs for distributed energy resources are the lower bounds for the price that sellers would accept. On the other hand, P2P trading with other market players is another good option.

In this sense a digital currency can be assigned for prosumers in smart grids to enable them to trade their load and generation. On the other hand, prosumers with various appliances cause huge load, generation, and flexibility forecasting uncertainties in the distribution networks, although controllable loads can use to handle this problem and minimize negative effects on power system. Moreover, with high penetration of prosumers in the low-voltage (LV) network, some barriers and limitations in terms of communication and network power flow appeared, which should be addressed by new technologies and management schemes. In addition, enabling business and market models is needed, which are compatible with new management schemes and operation principles. The target is to enable also participation of local small-size customers on providing different flexibility services. It is stated in literature that blockchain can provide innovative trading platforms where prosumers can exchange the surplus production and flexible demands [7]. These transaction can be performed in a secure way with transparent smart contracts [3, 8].

Prosumers can be considered individually in this market framework and make the transaction by their own. In this sense the P2P transaction among prosumers would be like Fig. 5.2.

Accordingly the generation and demand are measured by smart meters. The extra generation and demand are defined and transferred from one prosumer to another through some smart contract in blockchain platform. Based on smart technologies the load pattern and generation of RESs for prosumers are forecasted, and each prosumer would share the needs such as buying or selling the power within the blockchain platform in near real time, and the transaction among matched prosumers would take place. The market and transactions are performed virtually in blockchain, and the electricity would exchange through physical network as it is depicted in Fig. 5.2.

On the other hand, prosumers can be introduced and formed within different frameworks such as virtual power plants, microgrids, or aggregators that offer different values to the prosumers in the market [9]. It means that prosumers usually interact in the network as a part of a larger section of the network. Therefore the study of the prosumers has been conducted while considering one of these bigger entities. The selection of the entity is dependent on the placement of the prosumer. In the other words the prosumers who are a part of virtual power plants (VPPs) should be studied with the role of VPP. That will be the same

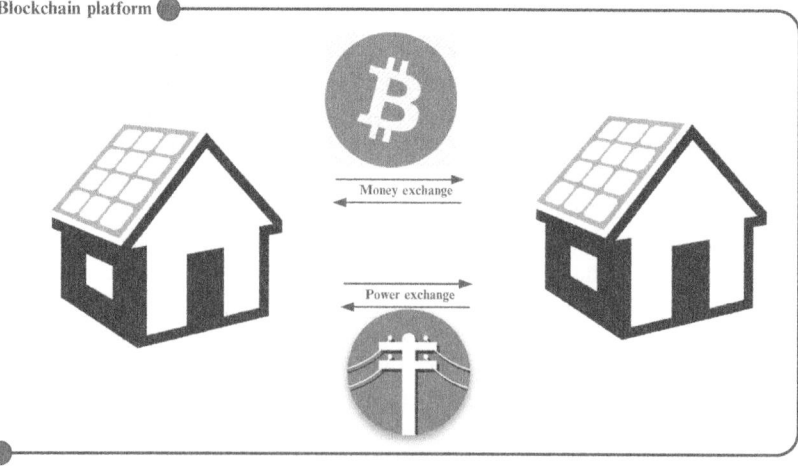

FIG. 5.2 Transaction among individual prosumers through blockchain platform.

for prosumers in microgrids or integrated by aggregators. It is noteworthy that there are also some independent prosumers that obviously can be studied in blockchain framework in a different way.

2.2 Aggregators

Aggregators serve as a broker for transactions between energy suppliers and several houses. These aggregators can be utility companies, commercial aggregators, commercial aggregators, or community groups who enable prosumers and customers to participate and transact in blockchain market scheme and platform [10]. Indeed, they are considered as validators for efficient use of DERs and prosumers while acting as a single entity. In blockchain scheme, it is possible that either DERs act individually or through aggregators in the market [11].

The better coordination among aggregators, TSOs, and DSOs provides a better solution for grid congestion management. Therefore blockchain developers are trying to find innovative solutions for automation and decentralized grid control. Since transaction speed for real-time grid management is also vital, suitable metering, grid infrastructure, control and communication systems in power networks, and among aggregators in this platform should be provided [7].

In a blockchain platform the transactive infrastructure has three layers as described in Fig. 5.3. The first layer is aggregators' data center where the electricity exchange is carried out. This layer is a virtualized set of servers operated by aggregators with a digital communication. The second is communication layer including all components requiring transactive controllers. The third is user layer where the transaction among different users through the IT infrastructure is performed.

The aggregators can be for VPP to deliver some services to TSO in ancillary service market. In this way, VPP aggregator is an interface among internal prosumers and external parties like TSOs, DSOs, and market operators [7]. The aggregator here refers to VPP owner or community manager.

A framework can be introduced in a way that aggregators of VPP provide generation or demand schedule for the relative prosumers while considering the network constraints. Through this schedule, aggregator can participate in ancillary service market effectively. It is supposed that the prosumers of a VPP are

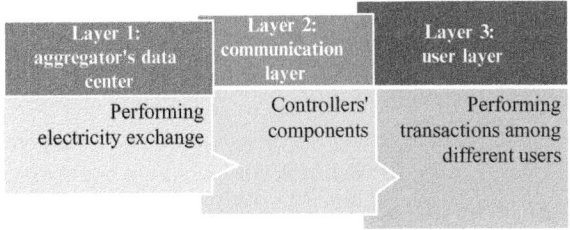

FIG. 5.3 Different layers in transactive controller of aggregator.

connected to the same distribution network and the aggregator schedule would occur while meeting the technical distribution constraints. In case of any deviation from the schedule, aggregators bid/offer the relative prosumers to buy/sell the required electricity in a special price via a smart contract within blockchain platform. Prosumers are able to react to the aggregator's bid in an auction by their transactive controllers. This smart contract is a set of rules encoded in blockchain to run the auction and define the accepted bids or offers for aggregators. Therefore, in the smart contract in the format of blockchain, after making bids by aggregators, prosumers with lower offers are selected until the required quantity of the bid is reached. The complete procedure is shown in Fig. 5.4.

Based on the Fig. 5.4, in a blockchain framework, the interaction of aggregators in the market can be formed in such a schematic way. Accordingly, aggregators can register their bids in the auction based on a smart contract while being lack of enough energy. At the same time, prosumers are able to register their offers. In this way, aggregators can integrate several prosumers, DERs, or other kinds of players who are able to compensate the shortage of power, although, in this example, it is supposed that prosumers are integrated by aggregators. These bids/offers are registered based on a smart contract mentioning the price and quantity of energy. Accordingly the prosumers whose offered price is lower than the aggregated bid will be selected. The blue line in the figure is all bids by aggregator, and the green bar lines are offers by different prosumers. As can be seen the offers by consumers with higher price than aggregator bid are not accepted. The selection will keep up till the all power shortage by prosumers are covered.

2.3 Virtual power plants

To coordinate a vast number of DERs with different owners, the concept of VPP can be a solution while transacting among all self-organizing prosumers. Indeed, VPP collects several numbers of coordinated DERs to have controllability, visibility, and impact at transmission grid. The concept aims to achieve upstream generation and transmission capacity reduction, network efficiency, and energy increment and pollution reduction. VPP operator is also in charge of providing upstream services to wholesale market and grid operators by aggregation of large number of prosumers and DERs [12].

The strategy to make a VPP varies based on the type of the incorporated DERs, the way of operation, and provided services. Accordingly, VPP can control the DERs directly, or this control can establish indirectly through sending incentive price signals to effect on prosumers' consumption and generation. In direct control, DERs can be dispatched according to their operating parameters and owners' preference [13]. In this case VPP has certainty over the facilities that DERs can provide like capacity and response. DERs in this control can provide fast timescale services such as frequency regulation [7]. To this end,

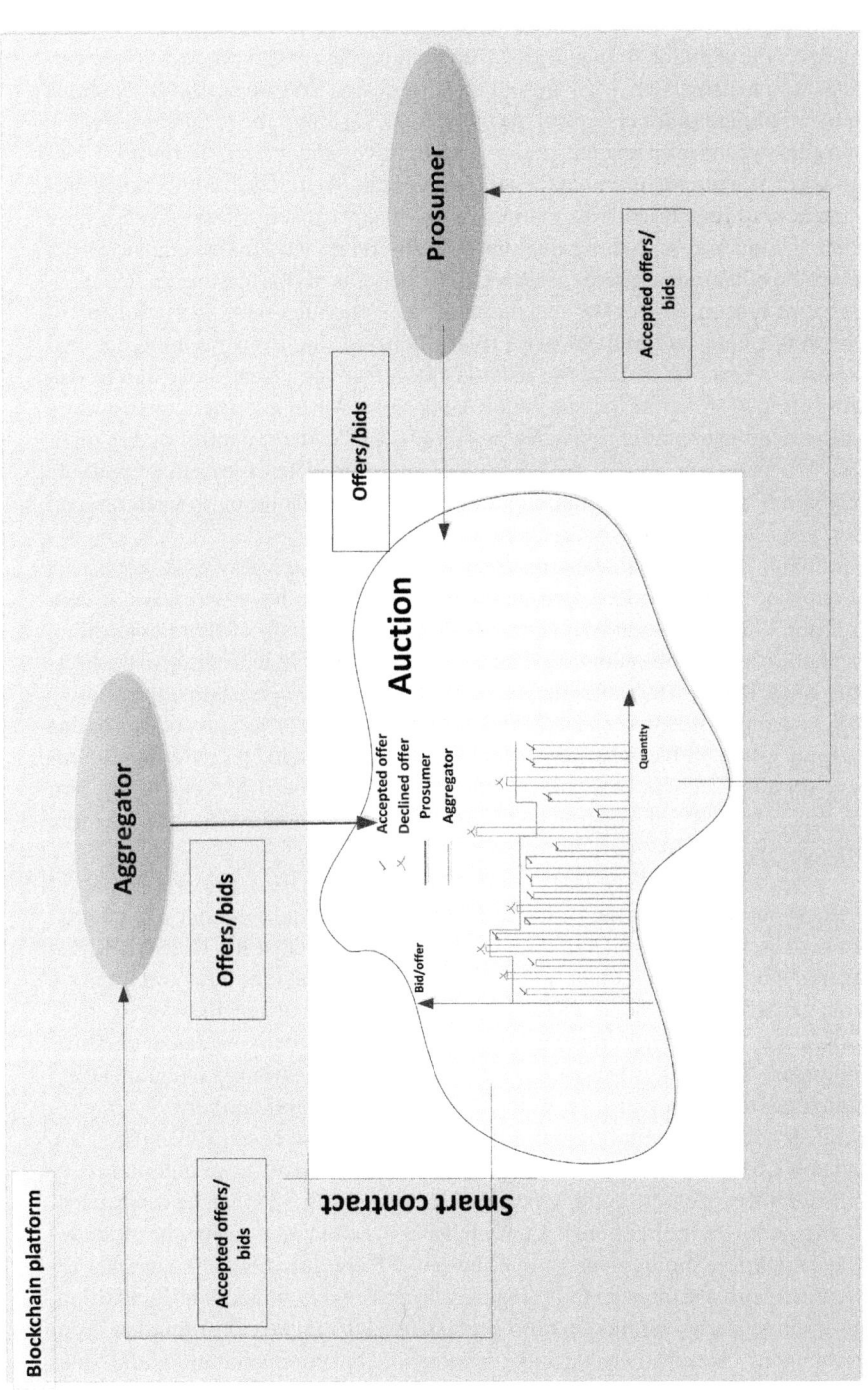

FIG. 5.4 Smart contract in an auction among aggregator and other players within blockchain platform.

distributed optimization methods are suggested such as Lagrangian relaxation run by VPP operator. It means that the communication and processing are carried out in distributed way, yet all require design and operation by a single entity. Through indirect control, prosumers decide about the local consumption and generation based on considering the incentives and their preferences. Time-of-use (ToU) pricing is one of the incentive pricing methods that encourage prosumers to shift the loads to reduce the upstream capacity. Day-ahead hourly pricing and location-based pricing in distribution network to coordinate DERs are some other examples of incentives. The benefits of this method are independence of prosumer over the scheduling of their flexible loads and reduction of communication requirement and privacy concern due to using unidirectional signals. On the other hand the method may cause new peak hour due to risk of shifting all loads to the special off-peak time. All in all, direct control does not provide enough flexibility for prosumers, and indirect control makes difficult for operator to predict the prosumers' behavior. Therefore an intermediate solution between direct and indirect method can be a solution to overcome all obstacles for VPP.

In Fig. 5.5 the possible interaction among VPP and the players in the market is depicted. As mentioned, prosumers in this structure have two ways of controlling, direct and indirect. An approach containing both of these controlling methods can be applied for a blockchain platform. Within this platform the location of DERs in distribution network is important, in the sense that the relative DSO would be able to run an accurate power flow. One of the advantages of this strategy is that VPP can participate in wholesale market and procure some ancillary services such as frequency regulation and reserve. Blockchain platform provides detailed information for VPP to be participated in the market more effectively.

VPPs with controlling large number of DERs are able to provide some grid services including ancillary services such as reserve and frequency regulation through some transactions in wholesale market organized by TSOs [14]. The interaction among DSO and VPP is also vital to reduce the loss and improve voltage regulation. Since VPPs have knowledge about the location of their DERs, they can provide some grid services for DSO for management of distribution network [15]. In this case DSOs will be able to efficiently integrate DERs by managing the power flow of the distribution network, actively.

It is noteworthy that microgrids can be also operated as VPP [16]. They include DERs and loads that can operate as a part of network or autonomously in an island mode. The communication architecture for VPP can be centralized, distributed, or unidirectional. Centralized communication is for the situation that prosumers communicate with a central VPP coordination, although distributed communication uses P2P prosumer-to-prosumer connection. In addition, in unidirectional communication, prosumers only receive information from coordinator. According to Fig. 5.6 the most suitable communication infrastructure in VPP for blockchain implementation is distributed one.

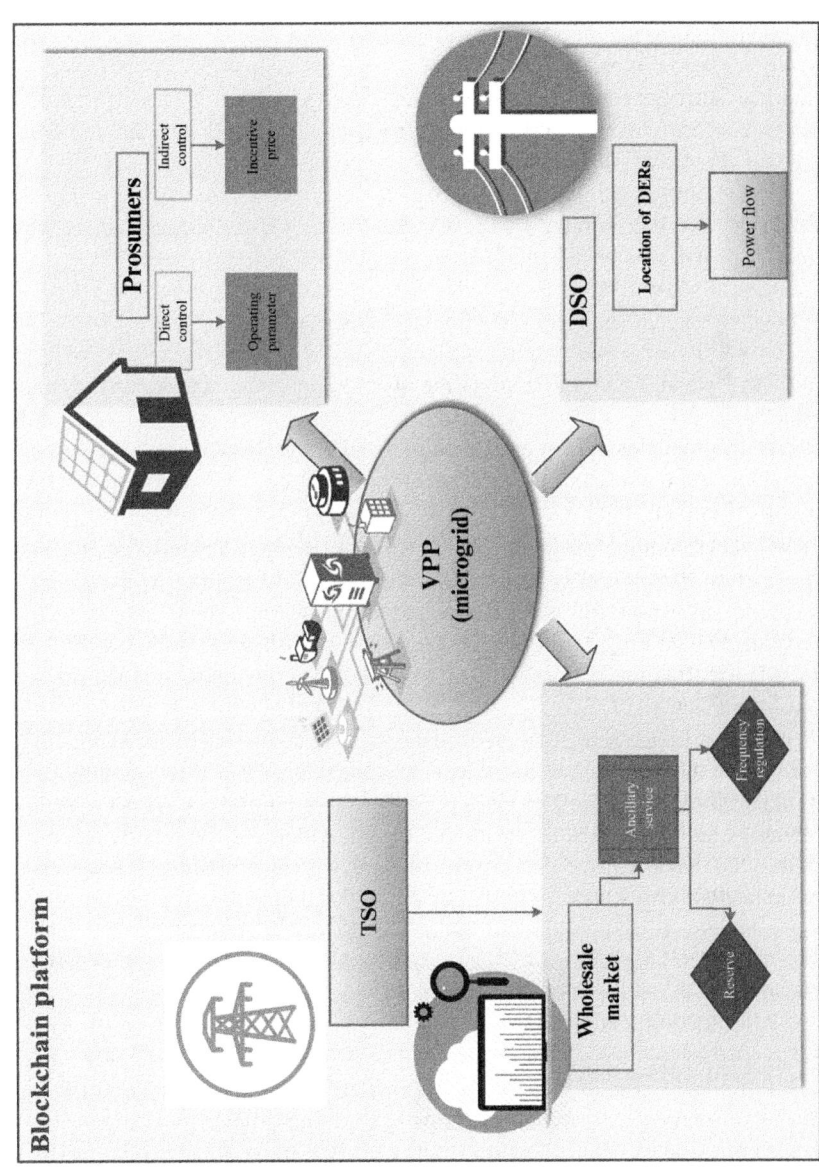

FIG. 5.5 Possible VPP architecture in blockchain platform.

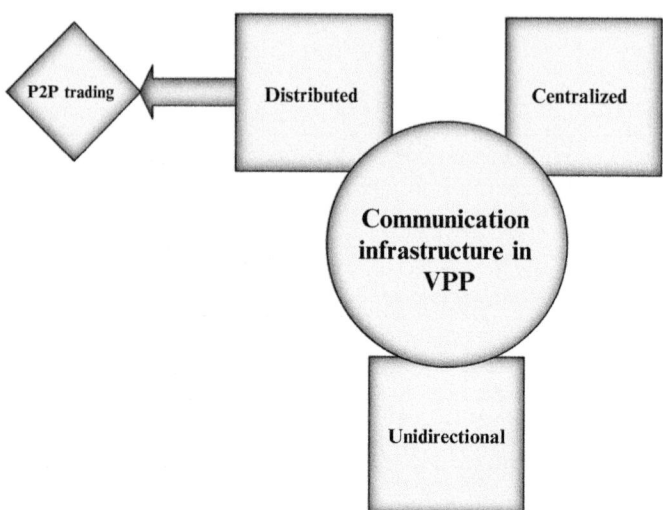

FIG. 5.6 VPP communication infrastructure.

2.4 Energy community manager

An energy community manager (ECM) aims to manage business activities within a community usually known as members with common interests and aims. Hence a community can be a microgrid, or a group of neighbors such as prosumers who are geographically close to each other. In addition, ECM plays also a role as intermediator among the relative community and the other parts of the market [17].

Within each community the sum of all the trades is controlled by ECM regardless which member is trading, because all the procedure is handled centrally by ECM. Therefore one of the objectives of ECM is to minimize the costs for each community subject to meet the balance constraint within the community among members. On the other hand the energy exchange with other markets outside the blockchain platform is also controlled by ECM. To this end, import and export energy is balanced, and the expenses for the communities to trade with other markets are aimed to be minimized. This model can be almost similar to day-ahead market model. The role of ECM is depicted in Fig. 5.7.

2.5 System operators

In blockchain-enabled systems, direct energy trading among local producers and consumers takes place. Therefore this energy trading is normally carried out in small communities. Nevertheless, the main question will be how to fit the mechanism to existing system operator companies such as DSO, TSOs, and ISOs who are eventually responsible for grid infrastructure and power delivery. Operators play a key role in the stable operation of the network even

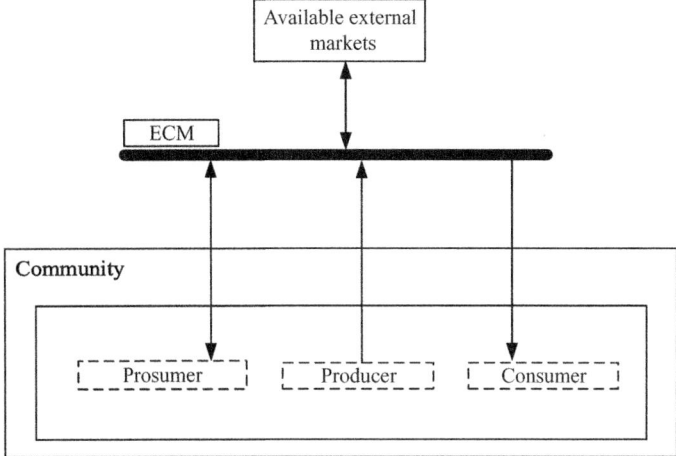

FIG. 5.7 ECM example.

when a complete decentralized operation is carried out in the network. The blockchain market can improve the efficiency of the market though [7].

Indeed, these operators make sure that the decentralized agreed contracts can be practically settled due to physical network constraints; hence the role of operators would be essential for implementation of blockchain. The benefits of blockchain market for operators can be as follows: First, blockchain can provide more precise information for operators in using the network and relative network fees such as distribution fees. Even for P2P transaction the grid charge should be accounted. Moreover, this recorded information about P2P transaction will help operators to better manage the capacity and power flow in the relative network. However, these potential benefits would need to have new management system to record blockchain information close to real time.

System operators' actions in blockchain platform should be limited because it may lead to market distortion and social welfare reduction. DSO has a great opportunity to increase the grid efficiency. To this end, one approach is to introduce a dynamic or nodal-based grid fees that work as an incentive for participants in the market. For example, prosumers should be more market oriented to participate in the market. Moreover, DSO can purchase different flexibilities such as reserve from flexibility market within the blockchain platform [18].

Regarding the interaction among blockchain and DSO, transition from complete physical electrical system to cyber system may cause some challenges about unfamiliarity and incompetence of DSO with new technologies. Therefore, if DSOs eventually will not be able to merge the cyber electric system, the ownership of the physical network should be separated from the operation. In this case the ownership of the blockchain platform would be for other players who will control all the end users' relations and access, and DSO just owns the cables [19].

Conventionally, network balancing was provided by TSO at high-voltage level, although nowadays, with increasing the number of RESs, the required flexibility for providing the balance can be applied within demand-side and low-voltage level. However, with large number of flexible recourses in this level, the complexity of operation would be higher for the operator. Therefore blockchain can be applied for decentralized operation of these flexible sources. This idea has been implemented by TenneT, a Dutch TSO, in cooperation with IBM as open-source blockchain provider and Vandebron as P2P trading platform in Netherlands and Germany [20]. Based on this project, EVs and customers' batteries are in charge of providing electric network balance to make the integration of RES more cost efficient. Accordingly, end users, through the electric cars and storage, provide flexibility for TSOs. There is no P2P transaction in this method, although a bidirectional communication between customers and TSO makes the transaction easier.

Blockchain is also able to remove the transmission and distribution charge for customers in which the former is about 12% and the latter is around 18% of the electricity bill [21]. In this sense, customers are able to freely transact with the generator rather than dealing with energy suppliers and grid operators. For example, large customers can purchase the electricity from PV farms or distribution network without the need of intermediary like transmission network. Therefore the transmission charge would be removed. Even in smaller scale, customers, with smart technologies in appliances, can purchase the required electricity from neighbors' PV. Therefore they can transact among themselves without need of distribution network, and its charge would be omitted.

To make all the aforementioned features practical, it is necessary to make the policies and regulations proper for DSO and TSO in all countries.

3. Blockchain business model

Blockchain technology was first employed as an effective mechanism for verifying cryptocurrencies, and later it was applied to broader economic issues such as energy transactions. Its application in energy markets was first proposed in [22] as a powerful tool for setting the value of RES based on the smart meter data.

Systems equipped with blockchain technology consist mostly of a distributed ledger, a decentralized consensus mechanism, and cryptographic security measures [23]. All users (participants or nodes in blockchain network) are allowed to directly share information and hold the copies of transaction records, called ledger. In the blockchain-based trading, the validity of a transaction is confirmed if all of the nodes achieve a common consensus [24]. Transactions should be confirmed through the use of predefined consensus in a shared execution manner, named smart contracts. Despite the unrestricted access of participants to transactions, each node can only access to transactions in which the participant was involved. Cryptographic hash functions are utilized so as to preserve privacy of inputs.

A set of rules is established in blockchain networks to approve the transactions. They are considered as consensus protocols. Today, proof of work (PoW) and proof of stake (PoS) are the most common consensus protocols [25]. All participants are required to do a complicated cryptographic puzzle so as to approve an energy transaction if the network consensus protocol is PoW [26]. Mining is the process through which the agreement is reached. This kind of protocol consumes a considerable amount of energy, time, and even money since they need to invest in the powerful hardware. In comparison, only nodes with highest stakes (stakes can be cryptocurrencies or values) can grant access to verify transactions, leading to a less energy footprint [27]. Reference [7] modifies the PoS protocol through the use of a hard-to-forge stake and a permissioned architecture. The modified protocol, named proof of energy (PoE), proved to be sufficient for energy markets as it aims at promoting self-consumption of peers in electricity markets, helping to reduce power losses.

Blockchain is categorized into private and public according to the nodes allowed to participate in the consensus process. In a public blockchain, anyone is able to join the blockchain-based network without any permission. On the other hand the accessibility options are limited for the participants of the private network. Moreover, only some designated nodes are chosen to verify transactions sharing in the private blockchain [28].

Some salient features of blockchain technology facilitate implementing a decentralized platform for energy markets. In a public blockchain-based market, there exists no central entity to manage, monitor, and approve the legibility of transactions. Accordingly, all of the users participating in the blockchain-based market have an equal right since they can submit a bid and accept an offer in a free and flexible way. Decentralization can also reduce the costs of transactions by eliminating the costs allocated for an intermediary. Besides, transactions recorded through all nodes are open and transparent, helping to solve problems related to information asymmetry [29]. As a result, it can guarantee the safety of the data information entered into the blockchain-based network. Finally the anonymity of counterparty is another key feature of utilizing blockchain so that participants can bilaterally trade with the other parties without knowing them.

Blockchain technology can be applied to wholesale, retail, and local markets. However, most of the studies analyze decentralization through blockchain in distribution networks and local markets. The most common distributed ledger technology (DLT) considered for energy markets is Ethereum, which can cope with tens of transactions per second [2]. However, Ethereum would not be suitable for a market consisting of many prosumers and consumers as it may take a considerable amount of time. Hyperledger, on the other hand, is able to support hundreds of transactions, making it a wise choice for a huge market [7].

The following sections discuss four main business models enhanced through the use of blockchain technology. Generally a business model shows the ways that the business works. To define a business model, some definitions should be introduced. Value in business models is defined as the amount of money paid to the seller [4]. A value proposition is the product offered to the customer [30]. Accordingly the value proposition defined in electricity market models can be electricity, ancillary services, and/or flexibility. The activities and resources deployed for distributing the value proposition are called the value chain [31]. In blockchain business models a value chain will be built on the platform connecting sellers and buyers of energy. The following sections introduced four blockchain-based business models including peer-to-peer (P2P), flexibility, OTC, and crowdsale trading platforms.

3.1 Peer-to-peer energy-trading platform

The recent development of information and communication technology (ICT) and new interests in managing demand and schedulable loads along with the policies on promoting distributed renewable resources have made small consumers play a proactive role of prosumers [17, 32]. These small prosumers seem to be willing to take part in electricity markets and change their roles from submissive ratepayers to the players who can schedule their resources and make profits.

However, the capacities of these small prosumers are mainly negligible, preventing them from taking part in a wholesale market and compete against conventional suppliers with huge capacities. These players are proposed to be aggregated by an aggregator. The aggregator as an intermediary would be responsible for aggregating these capacities, building offered bids on behalf of these players [33]. Although this condition can motivate prosumers to respond to the market prices and schedule their available resources more than before, they still do not have control over their offered prices and capacities. Besides, they should share the profit with the aggregator playing the role of an intermediary.

Local markets based on P2P trading give small prosumers the opportunity to make their own bids, schedule the resources, and manage their consumption. P2P trading enables the bidirectional flow of power and information in power systems while traditionally power flows in a unidirectional manner, from generators to end users. This kind of trading has a decentralized structure where all peers can participate in a pool-based local market or trade electricity or related services bilaterally.

A transition toward decentralization through local markets brings some benefits for the whole grid, local communities, and small end users. First, local markets (in a shape of virtual or physical microgrids) can keep a balance between supply and demand within their community reducing the burden on both transmission and distribution grids. Thus it would contribute to enhancing the reliability and resilience of the whole system. Power losses can also reduce owing to decentralization and self-consumption promoting through this kind of transaction.

Second, as aforesaid, regardless of the transaction costs, it reduces the costs for both suppliers and buyers participating in the local market by skipping the intermediary. Blockchain technology facilitates the aim of disintermediation by utilizing distributed ledger technology.

Moreover, P2P trading offers customers better choices for the source of energy they will receive. Suppliers are also given a chance to manage their own flexibility resources and maximize the profits. Besides, each transaction is allowed to have different prices taking into account the peers' preferences. Decentralization, therefore, is considered to promote democracy in local energy markets.

Consumer-centric electricity markets can make a major contribution to decarbonization as they encourage small customers to invest in renewables and make profits by managing their consumptions. As a consequence, local markets may empower communities to supply their own demand, making the communities autonomous.

Furthermore, P2P trading would mitigate market power in electricity markets. One of the leading reasons behind market power is the low elasticity of demand in electricity markets since the demand was traditionally unable to respond to the market prices. End users were mainly submissive ratepayers who were not subjected to a variation in market prices. Consequently, suppliers could easily exercise market power using financial withholding. However, in the local markets, players can react to market prices through their flexible resources (like batteries and EVs) and manage their consumption by flexible loads. Another reason resulting in market power is a limited number of players. P2P trading leads to all end users taking part in the local markets to buy and sell electricity. Hence, it enhances the liquidity of the market, heightens the competition between peers, and avoids monopolistic behavior.

One of the determining factors that is still considered as a big challenge for P2P trading is the grid operation [17]. In fact, there exist a few studies aiming at assessing the effects of P2P trading on the distribution and transmission grids. Although it was claimed that a P2P transaction leads to less power losses and congestions in both transmission and distribution networks owing to the short distances between suppliers and demand, bidirectional flows of power can pose a threat to the operation of the grid. So the transactions are proposed to be checked by the system operator before the energy exchange time period in case of any constraint violation [16]. An inspiring work concerning the impact of business models on grid operation could be found in [34] in which peers can contribute to providing flexibility services so as to help grid operation. The cooperation between distribution system operator (DSO) and transmission system operator (TSO) is another challenging issue that can be complicated by promoting P2P transactions.

3.1.1 Target participants

In general the main participants of local markets can be energy suppliers or sellers, energy consumers, or buyers and mediators [35]. Target participants are quite different according to the various designs for P2P markets. Small prosumers and consumers aggregated by another entity can be regarded as buyers or sellers in different market design. Considering the degree of decentralization and topology, there exist full, community-based, and hybrid P2P markets [17, 36, 37].

- In the full P2P trading, small players can individually participate in a pool-based local market or trade electricity bilaterally with another peer without any centralized supervision. Fig. 5.8 illustrates an example of a full bilateral P2P market. Possible interactions among actors are also specified in this figure. A multibilateral economic dispatch is also another form of full P2P market proposed in [38]. Participants of full P2P markets are end users playing the roles of prosumers and consumers. In this market, suppliers are small prosumers selling their extra electricity to the market. A prosumer can be any end user who owns one or several types of distributed energy resources (DER) including renewable energy resources (RES), electric vehicles (EVs), and batteries. Consumers are considered to be different kinds of end users who are not able to meet its demand by itself. However, a consumer would be able to manage its consumption according to the market prices to minimize the costs. The full P2P structure eliminates mediator, leading to the full decentralization.

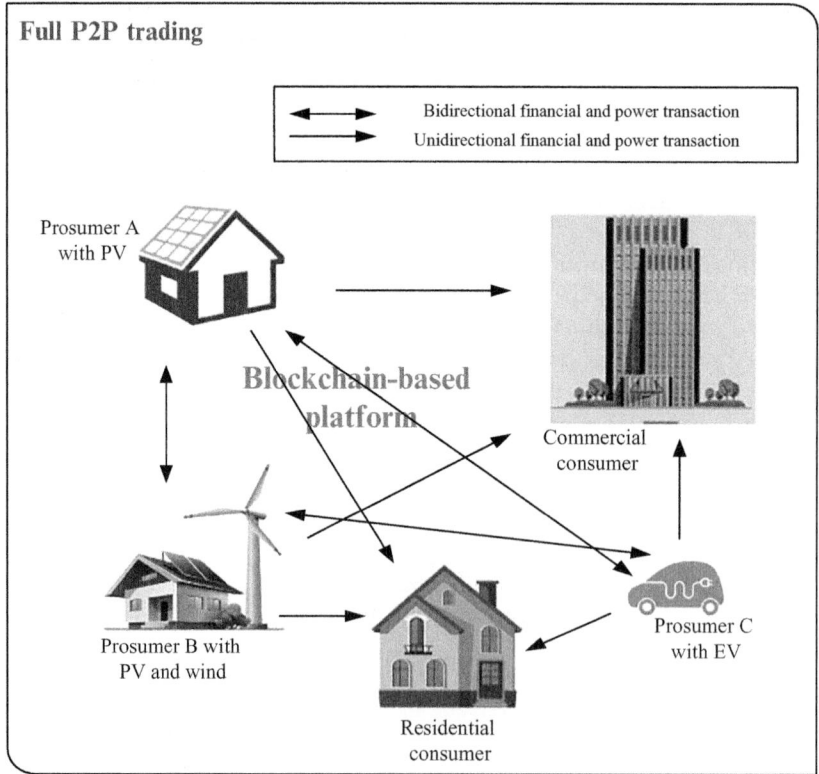

FIG. 5.8   A sample of full P2P trading.

- Community-based P2P markets consist of two types of mediator including a community manager and an intermediary connecting communities with the rest of the system. These two brokers act as the main seller and buyer. Small prosumers and consumers form a community due to geographical reasons or sharing common interests and goals [24, 25]. For instance this community can be a group of neighboring prosumers who want to trade energy with each other and some peers forming a community because they are willing to share green energy, although they are not at the same location [14]. Prosumers and consumers share their energy with a community manager. A community manager can be a kind of VPP if it integrates prosumers. Fig. 5.9 shows

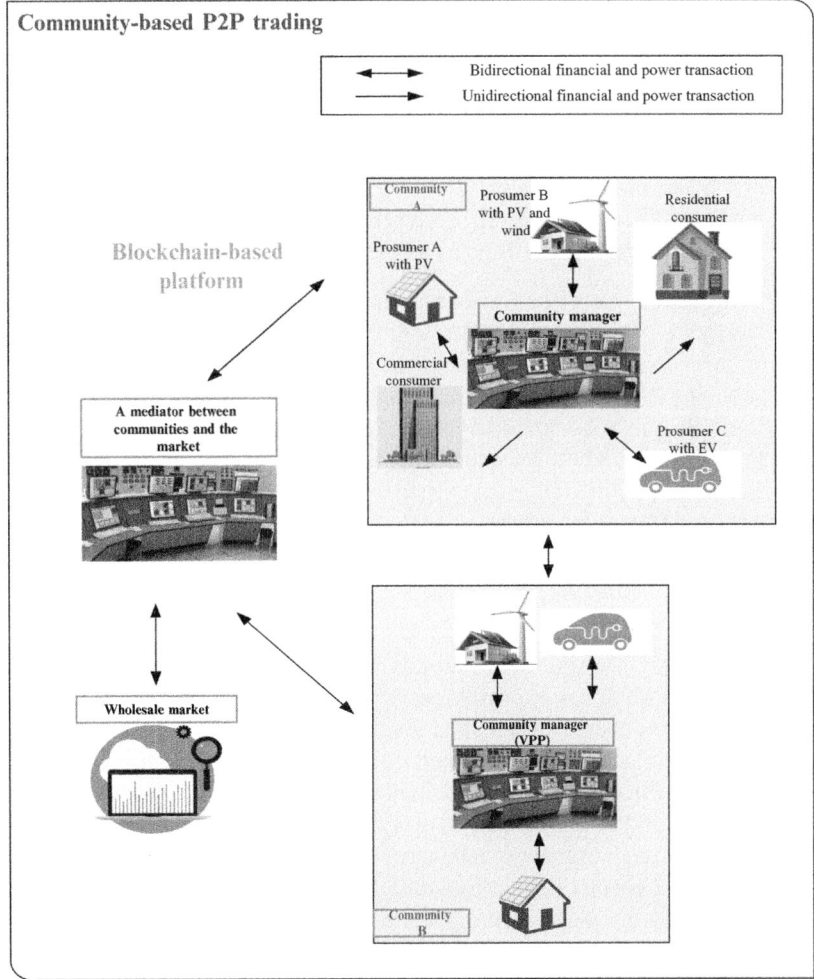

FIG. 5.9   A sample of the community-based P2P trading.

an example of a community-based P2P market with its participants and possible interactions between them.

- Hybrid P2P market: A hybrid P2P trading is a combination of both community and full P2P trading. In other words, prosumers can exchange the power individually or in the framework of the community/aggregator/VPP. The main actors of a hybrid P2P market are both energy communities and prosumers and consumers sharing energy with each other. Energy community, itself, can play the role of a mediator collecting some prosumers (forming VPP) or/and consumers to maximize the profit of these players. Reference [39] has offered small sharing zones in which they gather some prosumers so as to share energy with each other. A virtual entity called energy-sharing provider was proposed to act as a mediator who is in charge of coordinating the sharing activities within the community. Within this hybrid P2P structure, participants are prosumers, consumers, and an energy-sharing provider as a broker. The energy-sharing provider itself can act as a buyer or seller connecting a sharing zone with the other zones. A hybrid structure is an appropriate model as the preferences of all kind of end users can be considered in the model. Some small prosumers and consumers are not willing to participate solely in the markets as they do not have enough time to schedule their resources and consumption. An aggregator is assumed to take the responsibility of these peers. Fig. 5.10 shows an example of this kind of trading.

It should be noted that system operators including DSO and TSO play a vital role in all types of P2P market. Constraints associated with both distribution and transmission grids should be checked after each transaction so that they should not be violated.

Besides, prosumers mentioned in the models are not limited to the households equipped with RES. They can be an EV owner who wants to participate solely in a local P2P market, a battery storage or any player owning resources, and is able to bid to the market.

3.1.2 Roles of market participants

Roles of market participants vary according to the structure of P2P markets. The value chain of this business model unfolds on a platform through the use of private blockchain. Each participant is given a unique address in the blockchain. Participants submit their bids through their user accounts in the blockchain. The orders are settled, and the account balances connected to the users' addresses will be updated. Finally, miners verify transactions and generate new blocks [3].

While, in a full P2P market, small prosumers and consumers are allowed to submit offers through the blockchain network, only mediators are able to share energy between each other in a community-based P2P market. In fact the information related to the generation and consumption of small end users can be transmitted to the aggregators using smart meters and ICT technology. In a blockchain-based version, communities can trade energy with each other through the blockchain network without the help of any broker.

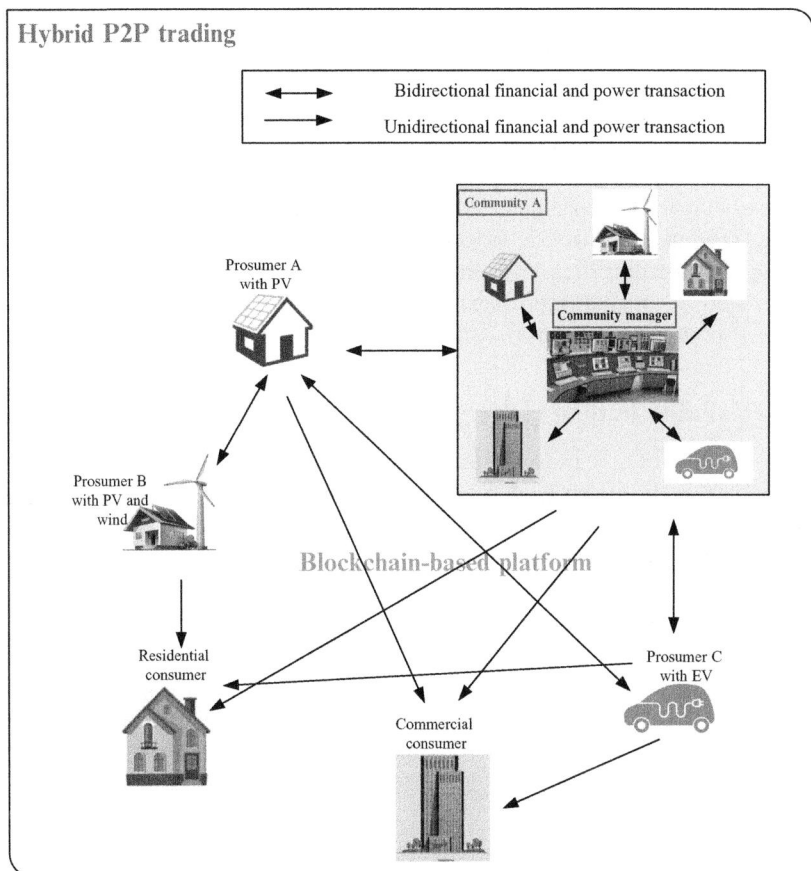

FIG. 5.10 A sample of hybrid P2P trading.

In a hybrid technology, however, consumers and prosumers as well as aggregators can grant access to the blockchain network and submit bids so as to gain profits. As previously mentioned, end users can also decide to play a nonactive role, being aggregated by another entity.

P2P trading enables prosumers and customers to schedule the energy resources and its flexible resources aiming at maximizing the profits or minimizing the costs. Following energy-trading purposes, DGs may be scheduled through connecting or disconnecting operations. However, reference [16] proposed that it would be better if the prosumer keeps DGs connected and utilizes its maximum power owing to the low operational costs of these resources. The participant can control uncertainties stemming from renewables by providing flexibility for the P2P trading through the use of batteries, EVs, and flexible loads.

EV users are proposed to participate in the P2P vehicular trading system in [40]. In the literature a P2P trading system was employed during the business hours to reduce the effect of charging EVs at these timeslots. Drivers decide to sell energy to another peer providing that the battery degradation and the efficiency losses were taken into account. In other words, they should be paid above the maximum value of the off-peak tariffs. On the other hand, drivers will accept to receive energy from other vehicles if they pay less than the market price. EVs can also be utilized as flexibility sources so as to help consumers or prosumers minimize their cost.

Flexible loads are considered as other sources used by consumers and prosumers. They would be able to reduce their consumptions at peak hours through shifting and curtailing their loads. In a community-based P2P trading, an aggregator would be in charge of managing flexible loads considering a standard degree of household's comfort.

3.2 Flexibility trading platform

The prevalence of using fossil fuels in different industries gives birth to some environmental problems such as global warming. Adoption of renewables along with electrification is introduced as effective solutions. However, the huge penetration of renewables in both transmission and distribution grids results in serious challenges due to the intermittent characteristic of renewable resources such as solar and wind power. Besides, these days, electricity supersedes fossil fuels. For instance, the share of EVs is growing rapidly owing to their efficiency and environmental benefits. EVs, themselves, can cause serious problems for the electric grid since it may change net demand patterns. The charging and discharging of EVs can also be quite unpredictable as long as they are not aggregated in the power systems. In addition to uncertainties coming from the new green resources, other issues may happen for distribution and transmission grids. Since small end users are encouraged to be equipped with renewables, reverse power flows may occur, making new congestion and voltage issues.

Flexibility products are considered as one of the optimal solutions to the aforementioned problems. To date, conventional generators are in charge of providing flexibility and reserves for the grid. These flexibility products are traded in reserve and intraday markets [41]. However, new problems with the power grid cannot be resolved by conventional flexibility resources. Ramping capability of conventional generators is not mostly sufficient [42], making the operator utilizing spinning or frequency-related reserves to capture uncertainties coming from renewables. Hence, seeking additional flexibility resources is of the essence so as to maintain the security of the system in a predefined level.

Generally, flexibility can be defined as the ability or possibility of changing consumption or generation patterns to react to external signals (such as price signals) aiming to maintain the stability of the system in a cost-efficient way.

Flexibility resources provided at distribution level regarded as real-time flexibility products that can be employed in both distribution and transmission grids.

Distribution-based flexibility products would be able to integrate renewable generations, provide real-time reserves, and mitigate market prices during peak hours. Batteries, EVs, and demand response (DR) programs are regarded as the most common flexibility sources provided at the distribution grid. Electricity can be stored at off-peak hours and injected during peak hours through the use of battery systems and EVs. As a result, these resources contribute to grid stabilization and balancing. Moreover, flexible demand can be also curtailed or shifted to other hours during peak hours that also will strengthen the power grid.

Flexibility products can be identified as the following:

1. Balancing flexibility at transmission and distribution grids [43]: This kind of product is offered to the transmission system operator (TSO), aiming at keeping the balance between generation and consumption. The current reserve and intraday markets were designed so as to provide flexibility related to balancing. However, these markets do not allow small flexibility sources to participate in them. For example, households with batteries, EVs, and responsive loads may not be able to take part and offer bids in these kinds of markets. On the other hand, balancing flexibility at the distribution grid is mainly provided by integrating small flexibility resources. To utilize flexibility products at distribution grids, the complete coordination of DSO and TSO is needed so as to check whether a flexibility product in one grid does not cause problems for the other grid.
2. Flexibility for the distribution grid [43]: The kind of flexibility is provided through integrating flexibility resources at the distribution grid to keep local balancing, manage congestion, and reduce power losses.

3.2.1 Target participants

Traditionally the main participants of flexibility trading at the transmission grid are conventional generators offering ramping capacities to reserve markets. However, aggregators with different sizes may also decide to contribute to transmission-level flexibility markets. Besides, distributed flexibility resources including batteries, EVs, and customers with flexible loads can play flexibility roles in local flexibility markets. Thus the target participants will be completely different according to the types of network (transmission or distribution network).

Balancing products at the transmission level were proposed to trade in wholesale markets in which both energy and services can be traded and the optimal amount of energy and flexibility products will be cooptimized in this market. In [44], flexibility products provided for the transmission grid were treated as ancillary services. In some proposed markets, suppliers submit ramp products and the corresponding costs to the operator; the operator decides to use them considering their costs if needed for the purpose of load following. A flexibility market can be proposed for providing flexibility at the transmission grid where huge aggregators consisting of a number of flexibility sources at the distribution grid can participate in this market. Conventional generators can also contribute to the market and submit their ramping capacities as well. TSO is considered to be another player submitting the required flexibility to the market.

Different market structures have been proposed to provide balancing flexibility for the distribution grid [45]. The studies mainly propose a separate market in which small flexibility producers would be able to offer flexibility products to the TSO. Again, TSO will have an option to choose the flexibility products needed regarding their offered prices and capacities.

Flexibility products provided for the distribution grid can be submitted in local day-ahead and intraday markets. Aggregators of flexibility sources are assumed to offer their products and the corresponding offered prices to the market. On the other hand, DSO also submits its required flexibility. Eventually the market is settled, and the optimum amounts of flexibility will be determined as well. DSO then participates in the wholesale flexibility and RT markets so as to resolve remaining security issues [2, 6]. We propose local markets constituting at the distribution grid so as to provide flexibility for the local distribution grid. Like community-based P2P trading, both flexibility aggregators and small end users with flexibility resources can submit the flexibility bids. On the other side, DSO, TSO, and also small customers may submit their required flexibility products.

3.2.2   Roles of market participants

The roles of the seller would be played by huge aggregators aggregating flexibility resources located at distribution grid and conventional generators, while TSO is considered as a buyer.

In local flexibility markets, small suppliers owning flexibility sources and aggregators at the distribution grid are regarded as sellers, and TSO, DSO, and small customers can purchase flexibility products. Small customers are those end users with renewables who want to compensate for the uncertainties of their resources. The implementation of the blockchain-based market would be the same, in comparison with community-based P2P markets.

In blockchain-based flexibility markets, the value chain is built on a blockchain-based platform in a way that all flexibility resources are connected through the platform. The type of blockchain would be private, so just specified players would be able to take part in the markets. With the help of blockchain technology, flexibility sources (at both transmission and distribution grids) can directly sell their flexibility products to TSO or/and DSO without any mediator. The value is better to be delivered through tokenization. Smart meters connected to the flexibility resources can share information in a real-time manner. Accordingly the flexibility provided by flexibility resources at the distribution grid is regarded to be real time.

3.3 OTC trading platform

Over-the-counter (OTC) trading is a kind of bilateral marketplace that mainly takes place at the wholesale level. Buyers and sellers are able to trade bilaterally without the help of any intermediary. Traditionally, this kind of transaction was conducted through phone calls or websites. With the advent of blockchain technology, players can easily share their offers via the blockchain network while staying anonymous. In addition, the data shared in the network cannot be tampered easily.

OTC trading can be categorized into spot or short-term trading and future one. Blockchain technology would facilitate short-term OTC trading as it can settle transactions in real time. In future OTC trading, large-scale operators would be able to hedge against price fluctuations [46]. Therefore future OTC trading may alleviate the risk of not being supplied for huge customers and assure suppliers that their product will be sold in the market.

3.3.1 Target participants

As the OTC transactions are done at the wholesale level, the players should be large-scale ones. Retailers, huge customers, generating companies, and aggregators are the main participants of this market. They can share information with each other directly. As a result the transaction costs will become lower due to disintermediation.

3.3.2 Roles of market participants

In this market the buyers can be retailers and huge customers located at the transmission grid while generating companies and huge aggregators can play the role of sellers.

Like other markets, suppliers share their offers in the blockchain network. The buyers will accept the offer according to the offered prices and their needs. The platform is set on private blockchain in which just specified parties are allowed to participate in the market. Fig. 5.11 shows a simple market for OTC trading platform.

Note that security must be checked for each individual transaction. However, in a pool-based market, it would be checked once when all transactions are initiated.

3.4 Crowdsale trading platform

Crowdsale trading or crowdfunding is a kind of trading in which a huge number of players can take part in the market, regardless of their size, capacity, and location. The trading is mostly done aiming at promoting renewable resources in the world.

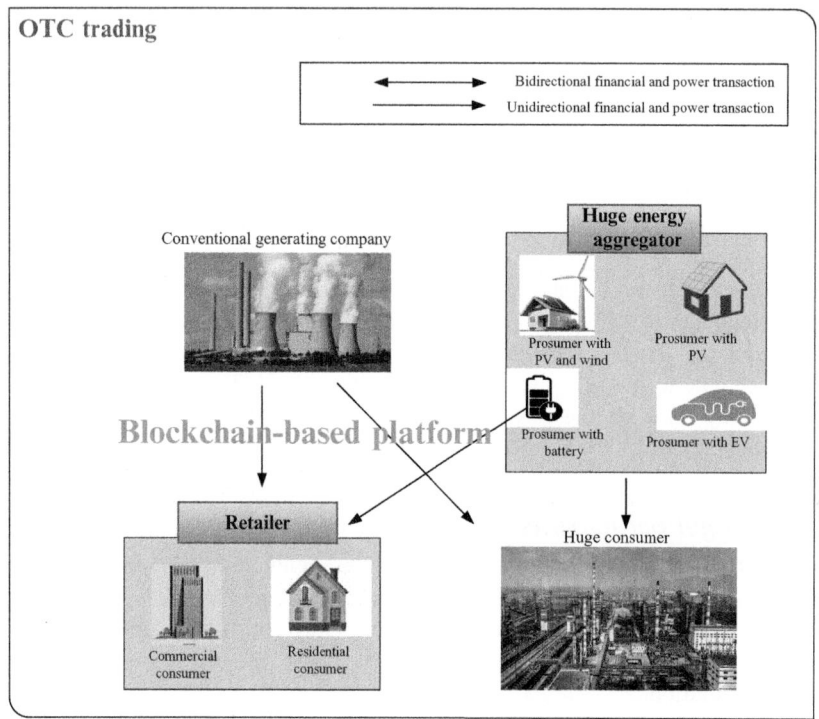

FIG. 5.11 A sample of OTC trading platform.

3.4.1 Target participants

Nonprofit organizations or generally private investors are players who want to invest in installing renewables (also called crowdfunder), while property owners are the main participants who are willing to install renewable sources on their private land.

3.4.2 Roles of market participants

There exist different business models related to crowdsale projects [47]:

In the equity-based model, the monetary profit of renewable resources would be shared between the property owners and the crowdfunders. In fact, investors offer the ratio of profit preferred to receive or the rent in exchange for investing in the green project, and the property owners accepting the offer can benefit from self-consumption and a part of profit coming from selling the surplus to the local markets. Property owners can also offer the rent paid to the investor, and the investors can decide to accept the offer.

In the reward-based business model, a property owner receives funds from an investor providing that the investor receives nonmonetary benefits such as a gift. The property owner will offer the benefit, and the crowdfunder can decide to accept it.

In the donation-based model, the investor donates the fund to provide the property owner with green energy. The donation is mostly received from wealthy organizations to poor developing countries.

Blockchain technology can eliminate the role of intermediary in the crowdsale trading. Accordingly, funders and property owners can easily trade with each other and share information via the blockchain network. The platform should be built on a public blockchain so that everyone can invest in renewables and every property owner is able to offer its land or a roof area for solar so as to be equipped with green sources. Note that the crowdsale market can be constituted globally. In other words an American organization may decide to invest in solar projects in Africa.

Crowdsale projects are considered as one of the best ways to promote renewables worldwide, leading to decarbonization. In addition, the global characteristic of crowdsale trading may help developing countries contribute to the reduction of greenhouse gasses. Besides the funds can be provided for very small projects that previously were neglected. Accordingly, everyone can benefit from green investment on their land.

4. Discussion

As a result, blockchain has provided a decentralized software platform enabling market participants to interact without any central entities or intermediaries. Accordingly, while running the market, blockchain platform evaluate the transaction based on predetermined policies. Therefore the roles for each market participants should be clearly and accurately defined. It should be noted that blockchain is not the only available solution for managing large numbers of (micro)transactions in P2P electricity markets, and there is strong competition from incumbent technologies that also enable very fast and secure transactions. This chapter, with classification of market participants in the blockchain framework, helps to manage and regulate the roles of market players in the future business models. Indeed, knowledge of current situation of different players in the available blockchain-based business model will help to improve the future structure.

Blockchain in electricity market is still facing many challenges in terms of implementation, fitting the available regulations and policies, and finding the best strategies and technologies to integrate all market players effectively that should be addressed by researchers. Moreover the lack of clear connection among the blockchain-based market in the lower-voltage level and upstream markets such as wholesale markets and retail markets makes the implementation of such a P2P business model more complicated. Finding a solution for this kind of interaction could move us one step forward to reach the comprehensive market scheme.

Acknowledgments

J.P.S. Catalão acknowledges the support by FEDER funds through COMPETE 2020 and by Portuguese funds through FCT, under POCI-01-0145-FEDER-029803 (02/SAICT/2017).

References

[1] K. Croman, et al., On scaling decentralized blockchains, in: Twentieth International Conference in Financial Cryptography and Data Security, February 22–26, 2016, pp. 106–125.

[2] G. Wood, Ethereum: a secure decentralised generalised transaction ledger, Ethereum Project Yellow Paper, 2014, p. 32.

[3] E. Mengelkamp, J. Gärttner, K. Rock, S. Kessler, L. Orsini, C. Weinhardt, Designing microgrid energy markets: a case study: the Brooklyn microgrid, Appl. Energy 210 (2018) 870–880.

[4] A. Orlov, P.M.H. Bjørndal, Blockchain in the Electricity Market: Identification and Analysis of Business Models (Master thesis), Norwegian School of Economics & HEC Paris Bergen/Jouy-en-Josas, Autumn (2017) 107.

[5] S. Wang, A. Taha, J. Wang, Blockchain-assisted crowdsourced energy systems, in: 2018 IEEE Power & Energy Society General Meeting (PESGM), 2018, pp. 1–5.

[6] S. Wang, A.F. Taha, J. Wang, K. Kvaternik, A. Hahn, Energy crowdsourcing and peer-to-peer energy trading in blockchain-enabled smart grids, IEEE Trans. Syst. Man Cybern. Syst. 19 (3) (2019) 1612–1623.

[7] P. Siano, G.D. Marco, A. Rolán, V. Loia, A survey and evaluation of the potentials of distributed ledger technology for peer-to-peer transactive energy exchanges in local energy markets, IEEE Syst. J. 13 (3) (2019) 3454–3466.

[8] M. Mylrea, S.N.G. Gourisetti, Blockchain for smart grid resilience: exchanging distributed energy at speed, scale and security, in: 2017 Resilience Week (RWS), 2017, pp. 18–23.

[9] S. Grijalva, M.U. Tariq, Prosumer-based smart grid architecture enables a flat, sustainable electricity industry, in: ISGT 2011, 2011, pp. 1–6.

[10] J. Mattila, et al., Industrial Blockchain Platforms: An Exercise in Use Case Development in the Energy Industry, 43, The Research Institute of the Finnish Economy, 2016.

[11] S. Chen, C.-C. Liu, From demand response to transactive energy: state of the art, J. Mod. Power Syst. Clean Energy 5 (1) (2017) 10–19.

[12] T. Morstyn, N. Farrell, S.J. Darby, M.D. McCulloch, Using peer-to-peer energy-trading platforms to incentivize prosumers to form federated power plants, Nat. Energy 3 (2) (2018) 94.

[13] J. Hu, G. Yang, K. Kok, Y. Xue, H.W. Bindner, Transactive control: a framework for operating power systems characterized by high penetration of distributed energy resources, J. Mod. Power Syst. Clean Energy 5 (3) (2017) 451–464.

[14] M. Andoni, et al., Blockchain technology in the energy sector: a systematic review of challenges and opportunities, Renew. Sust. Energ. Rev. 100 (2019) 143–174.

[15] N.Z. Aitzhan, D. Svetinovic, Security and privacy in decentralized energy trading through multi-signatures, blockchain and anonymous messaging streams, IEEE Trans. Dependable Secure Comput. 15 (5) (2018) 840–852.

[16] C. Zhang, J. Wu, Y. Zhou, M. Cheng, C. Long, Peer-to-peer energy trading in a microgrid, Appl. Energy 220 (2018) 1–12.

[17] T. Sousa, T. Soares, P. Pinson, F. Moret, T. Baroche, E. Sorin, Peer-to-peer and community-based markets: a comprehensive review, Renew. Sust. Energ. Rev. 104 (2019) 367–378.

[18] W. Cramer, C. Schmitt, M. Nobis, Design premises for local energy markets, in: Proceedings of the Ninth International Conference on Future Energy Systems—e-Energy '18, Karlsruhe, Germany, 2018, pp. 471–473.

[19] Energy News and Market Analysis (Ed.), Will blockchain disrupt the traditional distribution network model? Power Technology 08 October (2018).

[20] L. Diestelmeier, Changing power: shifting the role of electricity consumers with blockchain technology—policy implications for EU electricity law, Energy Policy 128 (2019) 189–196.

[21] "Smart Grids, Blockchain and the Changing Role of Transmission Operators | LinkedIn." [Online], 2017. Available: https://www.linkedin.com/pulse/smart-grids-blockchain-changing-role-transmission-alan-richards/. [Accessed: 5 July 2019].

[22] M. Mihaylov, S. Jurado, N. Avellana, K.V. Moffaert, I.M. de Abril, A. Nowé, NRGcoin: virtual currency for trading of renewable energy in smart grids, in: 11th International Conference on the European Energy Market (EEM14), 2014, pp. 1–6.

[23] E. Mengelkamp, B. Notheisen, C. Beer, D. Dauer, C. Weinhardt, A blockchain-based smart grid: towards sustainable local energy markets, Comput. Sci. Res. Dev. 33 (1) (2018) 207–214.

[24] F.S.B. Center, Consensus Methods in Blockchain Systems—Frankfurt School Blockchain Center, Medium, 9 July 2017. [Online]. Available: https://medium.com/@fsblockchain/consensus-methods-in-blockchain-systems-d2eae18b99b7. (Accessed 20 July 2019).

[25] A. Abidin, A. Aly, S. Cleemput, M.A. Mustafa, Secure and privacy-friendly local electricity trading and billing in smart grid, ArXiv180108354 Cs vol. 1, (2018) 1–13.

[26] K. Kvaternik, et al., Privacy-preserving platform for transactive energy systems, in: Proc. Middleware Conf., Las Vegas, NV, USA, December 2017, pp. 1–6.

[27] N. Courtois, On the longest chain rule and programmed self-destruction of crypto currencies, arXiv:1405.0534v11, December 2014 [Online], Available: https://arxiv.org/ (Accessed 11 February 2018).

[28] C. Natoli, V. Gramoli, The blockchain anomaly, IEEE 15th Int. Symp. Netw. Comput. Appl. NCA 2016 (2016) 310–317.

[29] C. Gao, Y. Ji, J. Wang, X. Sai, Application of blockchain technology in peer-to-peer transaction of photovoltaic power generation, in: 2018 2nd IEEE Advanced Information Management, Communicates, Electronic and Automation Control Conference (IMCEC), 2018, pp. 2289–2293.

[30] O. Gassmann, K. Frankenberger, M. Csik, Revolutionizing the business model, in: O. Gassmann, F. Schweitzer (Eds.), Management of the Fuzzy Front End of Innovation, Springer International Publishing, Cham, 2014, pp. 89–97.

[31] H. Chesbrough, The role of the business model in capturing value from innovation: evidence from Xerox Corporation's technology spin-off companies, Ind. Corp. Chang. 11 (3) (2002) 529–555.

[32] R. Zafar, A. Mahmood, S. Razzaq, W. Ali, U. Naeem, K. Shehzad, Prosumer based energy management and sharing in smart grid, Renew. Sust. Energ. Rev. 82 (2018) 1675–1684.

[33] H. Khajeh, A.A. Foroud, H. Firoozi, Robust bidding strategies and scheduling of a price-maker microgrid aggregator participating in a pool-based electricity market, IET Gener. Transm. Distrib. 13 (4) (2019) 468–477.

[34] A. Kargarian, et al., Toward distributed/decentralized DC optimal power flow implementation in future electric power systems, IEEE Trans. Smart Grid 9 (4) (2018) 2574–2594.

[35] M. Khorasany, Y. Mishra, G. Ledwich, Market framework for local energy trading: a review of potential designs and market clearing approaches, IET Gener. Transm. Distrib. 12 (2018) 5899–5908.

[36] H. Beitollahi and G. Deconinck, Peer-to-peer networks applied to power grid. In: Proceedings of the International Conference on Risks and Security of Internet and Systems (CRiSIS) in conjunction with the IEEE GIIS'07, 2007, 8 pp.

[37] Y. Parag, B.K. Sovacool, Electricity market design for the prosumer era, Nat. Energy 1 (4) (2016).

[38] E. Sorin, L. Bobo, P. Pinson, Consensus-based approach to peer-to-peer electricity markets with product differentiation, IEEE Trans. Power Syst. 34 (2) (2019) 994–1004.

[39] N. Liu, X. Yu, C. Wang, C. Li, L. Ma, J. Lei, Energy-sharing model with price-based demand response for microgrids of peer-to-peer prosumers, IEEE Trans. Power Syst. 32 (5) (2017) 3569–3583.

[40] R. Alvaro-Hermana, J. Fraile-Ardanuy, P.J. Zufiria, L. Knapen, D. Janssens, Peer to peer energy trading with electric vehicles, IEEE Intell. Transp. Syst. Mag. 8 (3) (Fall 2016) 33–44.

[41] L.J. de Vries, R. Verzijlbergh, Organizing flexibility: how to adapt market design to the growing demand for flexibility, in: 2015 12th International Conference on the European Energy Market (EEM), 2015, pp. 1–5.

[42] J. Villar, R. Bessa, M. Matos, Flexibility products and markets: literature review, Electr. Power Syst. Res. 154 (2018) 329–340.

[43] B. Zhang, M. Kezunovic, Impact on power system flexibility by electric vehicle participation in ramp market, IEEE Trans. Smart Grid 7 (3) (2016) 1285–1294.

[44] Flexible Ramping Product Revised Draft Final Proposal, Call 1/5/16, [Online]. Available: https://www.caiso.com/Documents/FlexibleRampingProductRevisedDraftFinalProposalCall1516.htm. (Accessed 21 July 2019).

[45] Y. Ding, S. Pineda, P. Nyeng, J. Østergaard, E.M. Larsen, Q. Wu, Real-time market concept architecture for EcoGrid EU—a prototype for European smart grids, IEEE Trans. Smart Grid 4 (4) (2013) 2006–2016.

[46] C.I. Dick, A. Praktiknjo, Blockchain technology and electricity wholesale markets: expert insights on potentials and challenges for OTC trading in Europe, Energies 12 (5) (2019) 832.

[47] D. Bonzanini, G. Giudici, A. Patrucco, (Chapter 21). The crowdfunding of renewable energy projects, in: V. Ramiah, G.N. Gregoriou (Eds.), Handbook of Environmental and Sustainable Finance, Academic Press, San Diego, 2016, pp. 429–444.

Chapter 6

Blockchain and its application fields in both power economy and demand side management

Ayşe Kübra Erenoğlu[a], İbrahim Şengör[b], Ozan Erdinç[a] and João P.S. Catalão[c]

[a]Yıldız Technical University, Istanbul, Turkey, [b]Katip Çelebi University, Izmir, Turkey, [c]Faculty of Engineering of the University of Porto and INESC TEC, Porto, Portugal

1 Introduction

In our modern world, technology, innovation, and digitalization all surround us and affect society from top to bottom. Remarkable opportunities and spectacular technological developments have great impact on the power industry in essence, recently. Therefore it is widely accepted that there is an urgent need to transform the traditional power system ushered in by Nicola Tesla some 120 years ago to the smart grid in new scientific horizon [1].

The traditional electrical grid has major limitations in terms of operation, management, and construction. In fact, this architecture was designed for fulfilling the needs set up in the last century [2]. It takes time to respond to dynamic changes in demand and/or generation due to applied vertical, multilevel [3], and centralized control mechanisms. Primarily radial construction and cybersecurity vulnerabilities cause to decrease reliability and resiliency parameters, which mean long-duration interruptions will occur in end-user services. Passive loads, in particular, are not controllable, and the system operator can only develop strategies in supply side to provide power balance at all times. In this structure the restricted energy storage system (ESS) potential has been evaluated as a significantly important issue in power system operation especially in case of any excess or shortfall in power [1]. Therefore it is not wrong to indicate that the capability of self-healing and self-restoration has been limited in the traditional grid without incorporating smart grid applications.

On the other hand, electrical energy is transferred from generally large-scale central energy plants established far away from the end users by enabling only one-way power and information flow [4]. Fossil-based resources (natural gas,

Blockchain-based Smart Grids. https://doi.org/10.1016/B978-0-12-817862-1.00006-3

coal, petroleum, etc.) have been utilized excessively in power system operation from its establishment up to now, which cause to increase greenhouse gas (GHG) emissions. In the light of recent events in climate, it is becoming extremely difficult to ignore GHG emission-based impacts on our ecosystem. Acid rains, ozone layer depletion, and increased carbon footprint are some of the most notable consequences that should be considered globally from government and legislative authorities' perspective.

To achieve key targets in terms of decarbonization of the power industry and combatting climate change, there are significant endeavors for deploying renewable energy sources (RESs) in the supply side. It is important to highlight that penetrating highest proportion of RES into the power grid could not be evaluated as an option; it has become an obligation [5] for providing a sustainable model. Besides environmental regulatory requirements, volatile energy prices of fossil-based resources and energy security issues also make energy transition necessary from nonrenewable sources to RESs [6]. Especially in case of any energy crisis, conventional systems would not be sufficient to meet increasing electricity demand.

Therefore, in the last few decades, photovoltaic (PV), wind, hydropower, biomass, and other renewable resource-based generation systems have attracted significant attention of the energy sector decision-makers and utility companies considering the aforementioned negative outcomes. The increasing investments have paved the way for growing this sector tremendously and accelerating technological innovations on this application fields thanks to the considerable amount of contributions. For example, according to United Nations' Sustainable Development Goal No. 7, it is aimed to increase the proportion of RESs in the global energy mix considerably in the year 2030 with the aim of supplying affordable and clean energy for all of us [7]. European Union has also made important attempts for expanding RES capacity by 27% in 2030, which is one of the most popular commitments that is accepted by all European countries [8]. It is important to indicate that the global average temperature can be kept between desired ranges if and only if renewable generation is increased by 23% for today to more than 50% in 2050 [5] with remarkable incentives. In fact the radical cost reductions in procured energy prices of PV and wind have important impacts on facilitating their integration into the power system and being the primary choice from the end-user side [9]. Thus 100% RES concept has gained importance in our modern world related to the aforementioned revolutions, and it has strongly been supported by governments and stakeholders.

The rapid transformations in the generation system have paved the way for increasing investments on a smaller scale, decentralized, and spatially dispersed systems instead of conventional large-scale centralized plants [10]. The main aim of this transition is to operate the sophisticated architecture efficiently with maximizing the reduction of both transmission and distribution power losses, reducing energy costs, and obtaining coordinated structure for future sustainable and resilient societies [11]. Therefore it would not be wrong to mark that high integration of utility-scale and domestic-scale RESs has a profound impact on the worldwide energy market that lies in the transactive energy concept, which gives incentives to all parties for trading energy based on the decentralized architecture [12]. These unprecedented changes in energy infrastructure and services have triggered to improve new strategies of grid operation, management, and new models in the context of commercial targets considering the reliability needs.

Unlike the all aforementioned advantageous issues, there are some important challenges to be taken into account, which are the intermittent, weather-dependent characteristics of RESs that seriously affect their power output patterns [13]. Therefore the vast majority of studies have been carried out to investigate the impacts of their dilute and disperse behavior on the power grid with a growing body of literature from the system operators' perspective. It is to be emphasized that such nondispatchable resources may cause supply-demand imbalances, voltage regulation problems, frequency instabilities, and other power quality disturbances in the electrical grid, due to their stochastic nature that should be handled [14]. Overall, these operational challenges that can affect the interconnected power system performance adversely highlight the need for incorporating different flexibility sources to maintain safety, robustness, and security. One of the top priority solutions is pointing out that the spinning reserve should be taken into consideration and become a necessary part of the operational tools. ESSs such as battery energy storage, flywheels, compressed air storage, pumped hydro, and superconducting magnetic sources [15] are one of the most fast-responding types of spinning reserve. Moreover, they can provide fast backup with responding load fluctuations in real-time and effective management strategies in case of any disturbances to smooth the overall voltage and power profile for maintaining normal network operation. On the other perspective the presence of ESS enables to reduce grid dependence and paves the way for performing decentralized mode in an optimal fashion with the objective of decreasing GHG emissions, energy, and operating costs and increasing power reliability [16].

Alternatively, electric vehicles (EVs), as one of the most promising storage technologies, are likely to become a key component for real-time applications with the ability to perform vehicle-to-grid and grid-to-vehicle modes [17]. It is obviously seen that the transportation system has also been transformed from fossil fuel-powered vehicles to emission-free, eco-friendly EVs due to the same concerns mentioned for generation side. From the power system planners' perspective, the highly penetrated EVs could play an important role in network operation based on the idea of backup storage or ancillary service provider to ensure supply-demand balance and meet the end-user requirements. These kinds of potential power sources will become increasingly popular in the future especially with the implementation of demand-side management strategies in the smart grid paradigm provided that technical, social, infrastructure, and policy challenges have been solved [15].

Apart from the mentioned ESS technologies, the concept of demand side management (DSM) has drawn significant attention and presents a promising solution for system operators to maintain voltage/frequency stability, defer generation plant construction, and increase energy efficiency [18]. Normally, energy generation increases in response to an increase in end-users' demand in traditional power system operation. However, ever-increasing demands and the widespread adoption of distributed generation enforced to change the mentality and demand side become new axiom area with the help of outstanding developments in the fields of smart grid technology. And DR is one of the most popular techniques of DSM that enables end users to reduce/shift their electricity consumption in response to electricity prices or operator's requests [19]. The communication and information technologies and an advanced metering infrastructure (AMI) enable two-way communication between the utility company and the end users for implementing load shifting/load reduction strategies by taking into account operational benefits in the controllable platform of smart decentralized structure [20].

Therefore all of these can be evaluated as a forerunner of modern power grid architecture, which is currently undergoing drastic changes in both supply and demand sides. In the smart grid era, it is expected that almost every end user can produce their own energy by installed on-site distributed generation units called as prosumer. With incorporating smart meter technologies, two-way communication will be available and provide information exchange between distribution system operator (DSO) and end users that presents significant advantages in terms of controllability, observability, security, and stability [21]. The end users can observe their production and consumption at the same time while it is possible to participate in DR programs depending on the real-time electricity prices broadcasted by DSO. All the aforementioned changes paved the way for unleashing a revolution in power system that requires operating in decentralized, peer-to-peer fashion instead of centralized methods.

Processing, storing, and supervising the huge amount of data have become one of the most challenging issues considering hundreds of millions of terminals of the utility grid. Also, there is a growing concern about cybersecurity vulnerabilities in case of any single-point failure due to the existence of only one control center in a centralized operation manner [22]. Furthermore, sophisticated and complex communication infrastructure required costly investments that do not have to meet our modern world necessities [23]. To address these issues, blockchain or distributed ledger technologies (DLT) were mainly presented to facilitate peer-to-peer decentralized transactions between nodes without requiring any third-party participation [24]. A better management and operation could be ensured by blockchain-based implementations and applications, reducing the system's dependency on the utility in a decentralized architecture.

2   Blockchain technology in different areas including power economy

This emerging modern world was driven by an unending stream of next-generation communication and information technologies, which are Internet of Things (IoT), cloud computing, and big data [25]. Similarly, blockchain is one of the greatest innovations among emerging technologies that would ultimately propel us into our modern age. In fact, it is the underlying technology of Bitcoin that was the major milestone in establishing a decentralized architecture. Bitcoin, the world's first cryptocurrency, was presented in 2008 by an unknown author or group of authors calling themselves as Satoshi Nakamoto [26]. The implementation of Bitcoin in 2009 triggered a huge amount of innovative scientific inquiry. Seemingly, this marks a crucial turning point facilitating peer-to-peer and distributed transactions in many fields of power economy.

Recently, utility decision-makers, startups, financial institutions, national governments, and the academic community have shown an increased interest in blockchain applications [24], and it has been comprehensively investigated from different points of view. Obviously, it is against today's traditional sophisticated financial payment architecture. In this system, any single point of failure causes undesirable consequences due to the vulnerability of both technical failures and malicious attacks. Also, to operate the system, transacting parties should trust central intermediary that introduces extra costs, time-consuming exchanges, inefficient concurrency control, and insecure data storage [21].

However, on the other hand, blockchain is a digital and distributed data structure that enables to make a transaction between two mutually unknown and otherwise unrelated parties directly without the existence of any trusted central authority [27] that sounds very new, creative, and innovative. Generally speaking, the system is operated by the created technical codes and rules that are determined by the community, that is, by the network users in the system independent of any legal financial authority or any regulatory body [28]. This is one of the oddest yet most brilliant core characteristics of blockchain technology. What remains unclear is how such an operation is possible considering the long-lasting effectiveness of the third parties.

In fact the banks are eliminated, and each individual network members (nodes or clients) become a new bank that is capable of storing the information of any digital transactions, records, and executables in their distributed ledgers as compared in Figs. 1 and 2. Each event is ordered chronologically and copied to every node in larger forms sequentially by attaching to previous blocks in chain form [24]. It is possible for network users to reach the blockchain and view its contents and participate in the consensus process as an active member [29]. A primary concern of this architecture for operating the system without the assistance of any third party is double spending and fraud. As we all know the cryptocurrency is digital that can be generated after executing particular

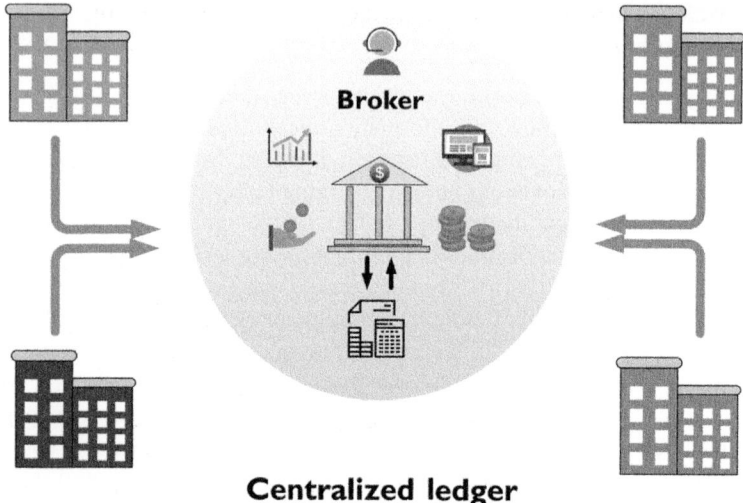

FIG. 1 The architecture of traditional centralized ledger system.

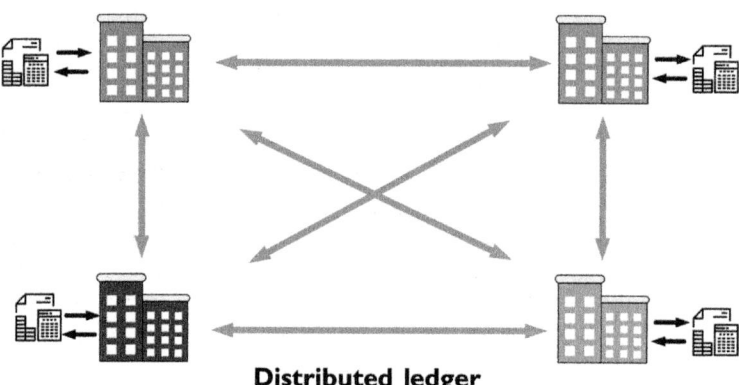

FIG. 2 The architecture of distributed ledger system.

cryptographic algorithms and protocols. So, it is important to prevent copying the same amount of currency on the individual network users' computer, mobile phone, or workstation wherever it is stored and sending them to the receiver point more than once.

To address the questions that have been raised about the security of the blockchain-based system operation, there is a simple yet clever concept that enables to keep users' identity anonymously with public-key cryptography [30], an asymmetric cryptography protocol, is used in this architecture. Two different cryptographic keys are provided for each user, namely, private key and public key comprising of numeric or alphanumeric characters [31].

The private key is randomly generated, and it is to be kept secret for users' own security that is used in signing their own transactions [32]. However, they could not be considered independent of each other, that is, there is a mathematical relationship between them enabling to generate from private key to the public key, while the reverse process is almost impossible thanks to the robust encryption codebase [31]. Therefore it can be indicated that there is no problem for sharing the public key with other participants, yet it makes the user addressable in the network. Users have been known by their digital signature combining private and public keys that makes the transaction extremely secure for all of the participants. The cryptography procedure is illustrated in Fig. 3. Basically, in this process, the payment message consisting of the recipient's public key, address, and amount of payment has been created by the sender and transferred to the receiver end accompanied with sender's digital signature securely. This cryptography provides an important opportunity by updating every transaction in the network participant's distributed ledgers and the organized time-stamped blocks even their offline copies are held in the blockchain, which shows its permanent and traceable architecture. So, this makes changing data or information irrevocably hard thanks to advanced cryptographic techniques [25].

On the other perspective, it is expected from network users to validate whether the transaction is performed appropriately or not for the purpose of building trust between participants without any central authority. However, there is an urgent need to prevent appending the data in blockchain whoever wants to do it. Otherwise the system will unable to withstand malicious attacks and encounter some important challenges to be handled. This mentioned issue has received considerable critical attention, and cryptographic tokens are presented as a solution for encouraging the honest nodes to add only executed transaction information on the system [29]. Widely known as miners can be identified as a major contributing participant in the network ensuring that false data cannot be inserted and time-stamped blocks cannot be tampered by any untrusted members making blockchain trustable, secure, and resilient.

When any user attempts to transfer a certain amount of digital currency from her/his electronic wallet to another user's, the transaction is to be verified by the network miners and formed them as "block" to integrate the chain. The basic process is illustrated in Fig. 4. A major problem with the confirmation process

FIG. 3 The illustration of cryptographic process.

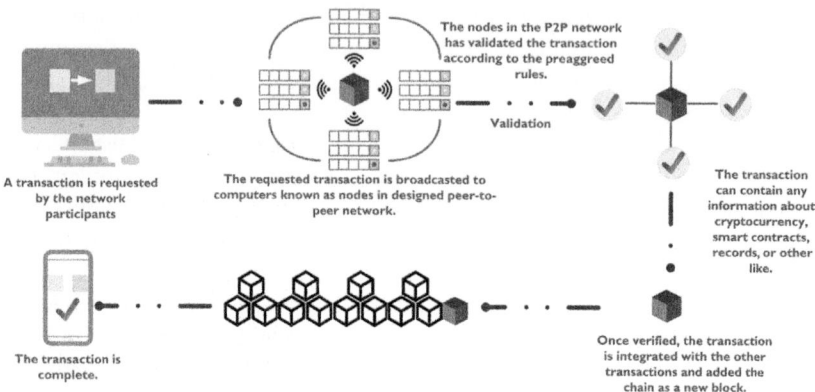

FIG. 4 The basic process of validated blockchain.

is finding the hash output of block, which is related to the stored information. It is worthy to note that using a specific cryptographic hash algorithm enhances the security of data drastically similar to the private and public keys. The illustrative example of hashing of blocks is depicted in Fig. 5.

The hash function converts the input in string form (numbers, letters, media files, and/or symbols) of any length into the fixed length of hash output (aka signature) that can vary (32-bit or 64-bit or 128-bit or 256-bit) based on the utilized hash function. For example, Bitcoin uses SHA-256 in its process for producing unique hash outputs [33]. To extend the knowledge, it is necessary to draw our attention to the distinctive characteristics of hashing cryptography that should be considered. Entering the same hash input results in creating the same hash output, that is, changing the input multiple times has any impact on output characters. Also, it is sensitive to any changes in input value; even only one letter or number means that entirely different hash output will be produced after a process. Moreover, one of the most impressive qualities of hashing is that one-way transformation is possible in this architecture, which is not possible to obtain the original data set from hash output [34]. Herein, miners have great endeavors to solve a cryptographic problem (finding a hash output) that will help to be rewarded with cryptographic tokens in return for appending the verified block into the chain. This complex mathematical problem, in particular, requires high CPU resources and a considerable amount of computational work from the miners' perspective that the process should continuously be repeated to reach the signature (hash output) requirements [35]. These all specifications of blockchain architecture make major and indeed an essential contribution to the field of achieving securely and reliable distributed transaction by implementing advanced cryptographic techniques.

The study made by UK Government Office for Science [36] makes a major contribution to research on combined blockchain technologies with smart

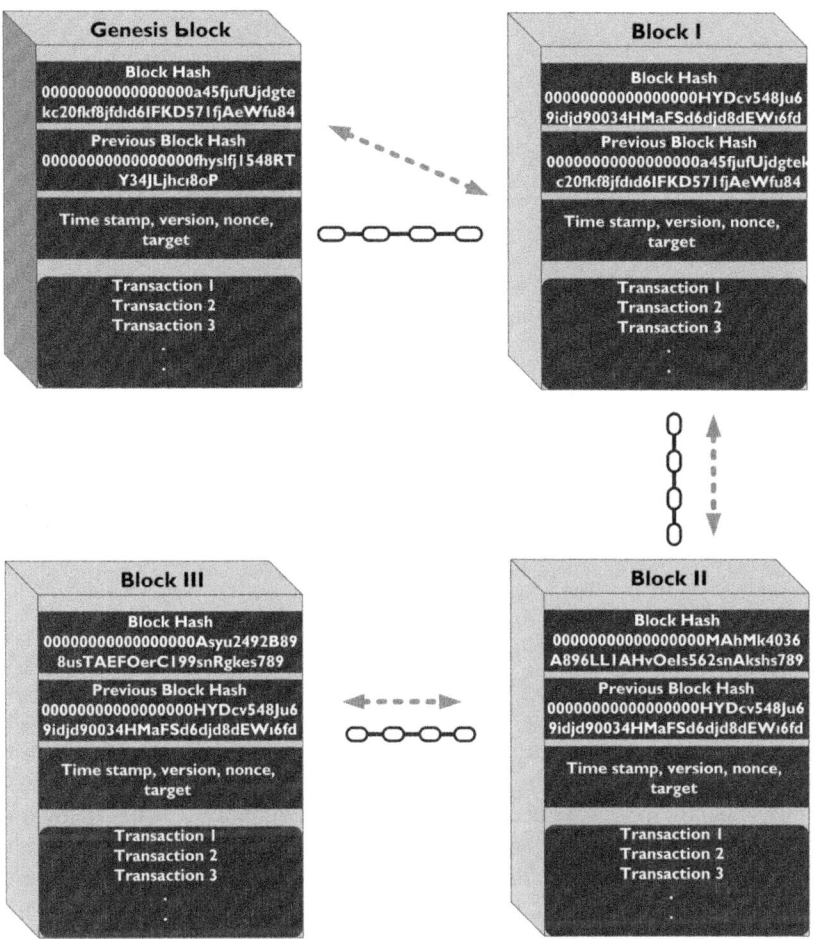

FIG. 5 The illustration of simple Bitcoin blockchain architecture.

contracts and provides an exciting opportunity to expand our knowledge of utilizing its potential as much as possible. The illustrative example is shown in Fig. 6.

There is great number of important areas where it is possible to improve novel business solutions by integrating smart contracts. The terms were first described in the 1990s by Nick Szabo, a notable computer scientist and cryptographer and who defined it as [37]: "a set of promises, specified in digital form, including protocols within which the parties perform on these promises." Although a smart contract presents a considerable amount of advantages and opportunities, it was not a convenient time to implement its real industry [38]. The emerging technologies, Bitcoin, and blockchain revitalized the smart

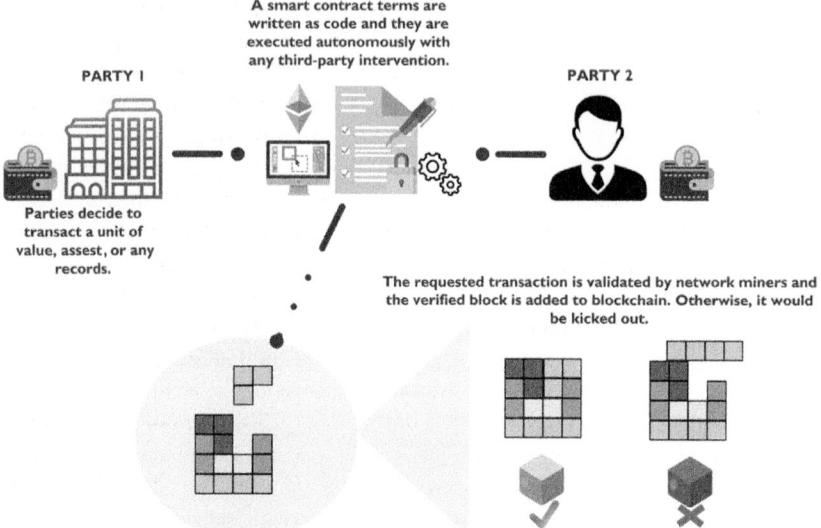

FIG. 6 The illustration of smart contract-based system operation.

contract paradigm and paved the way for developing new strategies in different application fields. Especially in 2015 the platform of Ethereum was released, which is one of the most useful things ever discovered for smart contracts and immediately leading participants to build their own distributed applications as "autonomous entities" [39].

Technically speaking a smart contract is a self-enforcing agreement between parties, mutually unknown and otherwise anonymous counterparts, which is capable of executing the rules automatically in satisfying certain conditions [40]. Needless to say, this interesting concept presents the decentralized architecture, removes the intermediaries (lawyer and/or other central agency), and determines its own rules with cryptographic code, which is completely different from a traditional contract framework. Within the blockchain context, the agreed terms are embedded in computer code written in solidity as one of the most brand-new programming languages [41] and are established as sophisticated if-then statements. In other words, this contract has been prepared to contain all regulations, details, and enforced programmed rules to run based on If-This-Then-That (IFTTT) logic in which the directives are executed sequentially [39].

A vending machine can be evaluated as a primitive example of smart contract in which the transaction is executed based on the encoding rules into a machine [42]. Anyone inserts the required amount of coin for purchasing what she/he wants to and presses the numbers related to that product. After, it is to be controlled by the machine similar to the smart contract whether the correct amount of fund is inserted or not. If the answer is yes, the product will be ejected together with the change (if extra money is inserted in the beginning of the process). Here's also a typical example of a smart contract for expanding our knowledge on this concept [43]: "let's say you and I have agreed that if I write you a history of bitcoin, you'll send me $10 on my birthday this year. We can do that via a legally enforceable contract, which involves lawyers, notaries, and so on — or we can do it via Ethereum. In the latter case, you put $10 worth of smart coins in escrow, and when the terms of the contract are met, those coins are released to me. If I don't meet the terms of our agreement, the coins are released back to you." So the instructions are executed automatically based on the script written beforehand if and only if predefined conditions are satisfied [39].

Overall the developmental stages of blockchain technology can be divided into three consecutive groups considering the intended audience. Virtual cryptocurrencies (e.g., Bitcoin) have emerged in Blockchain 1.0 that was the crucial turning point in terms of indicating to start the new digital era [44]. Blockchain 2.0 is seemingly accepted as the major milestone of enabling transaction beyond cryptocurrencies by executing smart contracts autonomously [32]. These two of them are the forerunner of the next-generation blockchain 3.0 in which it is possible to implement the distributed architecture in many fields, such as government, health, science, and IoT. The history of blockchain-related innovative technologies is illustrated in Fig. 7.

Therefore it would not be wrong to mark that blockchain is one of the greatest technological innovations that has paved the way for opening a range of new opportunities for both financial and nonfinancial areas. There is a great potential to change especially financial system entirely thanks to the peer-to-peer distributed structure; the critical issues can be handled by this foundational technology with transforming the paradigm from third trusted authority to trusted math [45].

Today, it is well known that from transportation systems to supply chain managements and energy trading implementations to communication, health services are operated by centralized points of control, which is not possible to think of real-time implementations without it. However, the presented innovative features enable blockchain to find many application fields in our modern world that have the potential to revolutionize society completely without a doubt. Marc Andreessen, the doyen of Silicon Valley's capitalists, has also indicated last year that the blockchain-distributed consensus model is the most important invention after the Internet itself. On the other perspective, Johann Palychata from BNP Paribas has pointed out that Bitcoin's blockchain and the software is such an inventions that will have great impacts on finance and beyond like the steam or combustion engine [46]. Thus the radical changes are expected in a vast range of fields from stakeholders by overcoming the challenges in an innovative way. It is important to combine emerging brand-new technologies (such as machine learning) with blockchain to improve the efficiency and current practices of systems and to accelerate the speed of services [47]. A growing body of literature studies and important pilot projects has been conducted with great contributions to extend our knowledge of this technology

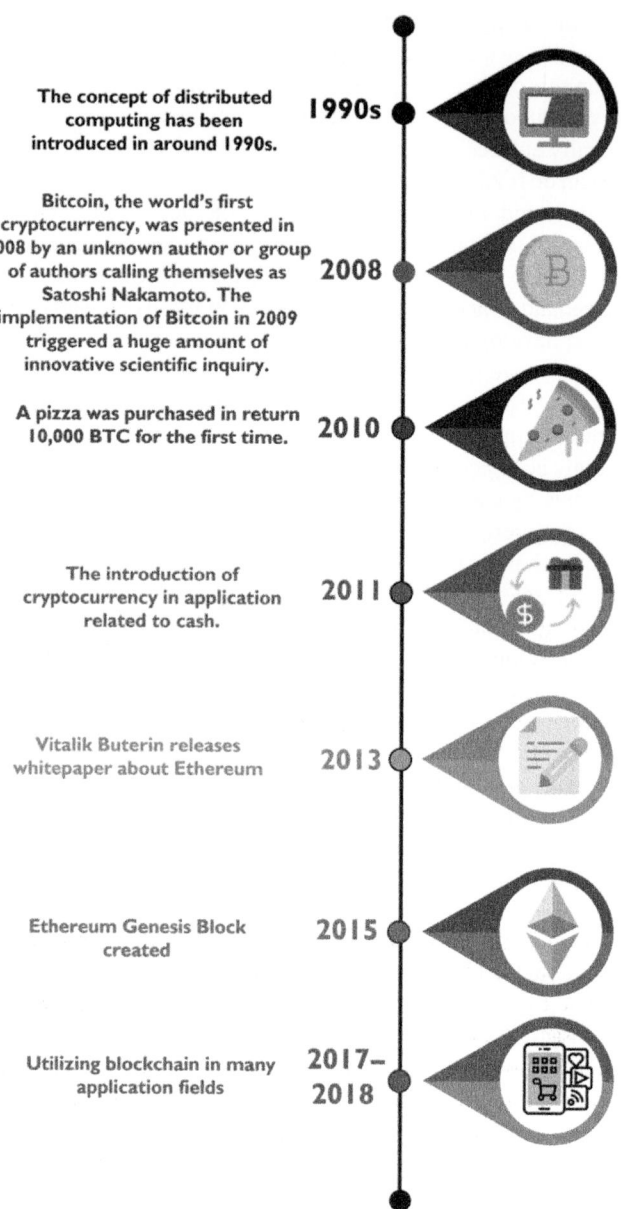

The concept of distributed computing has been introduced in around 1990s.

1990s

Bitcoin, the world's first cryptocurrency, was presented in 2008 by an unknown author or group of authors calling themselves as Satoshi Nakamoto. The implementation of Bitcoin in 2009 triggered a huge amount of innovative scientific inquiry.

2008

A pizza was purchased in return 10,000 BTC for the first time.

2010

The introduction of cryptocurrency in application related to cash.

2011

Vitalik Buterin releases whitepaper about Ethereum

2013

Ethereum Genesis Block created

2015

Utilizing blockchain in many application fields

2017–2018

FIG. 7 The history of blockchain-related technological innovations.

both in the field of theory and practice. Some of the most widely implemented areas are summarized in Fig. 8 and explained with examples in blockchain-based system architecture providing trusted interaction between participants and technology.

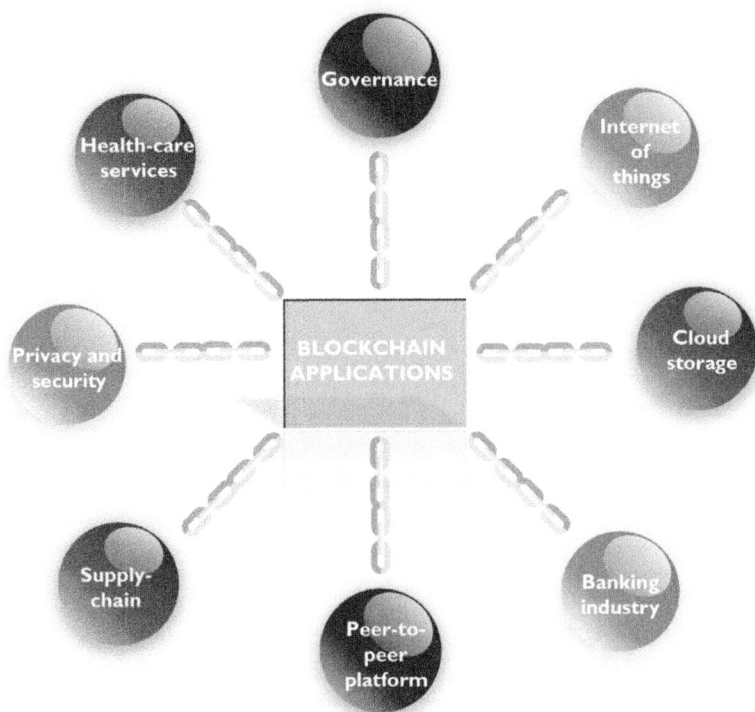

FIG. 8   Blockchain application fields.

As expected, the real adoption of blockchain technology will find a primary area in the financial industry that creates drastic changes in the system operation and enforces to replace major parts of their model. It is widely known that banking system is a huge network of integration of different organizations, which allow the transaction of digital assets, funds, or values among multiple parties (lenders, clients, capitalists, investors, etc.) [48]. The main drawback of the conventional banking system is that the processes are completely controlled by central authority, which makes protecting the customer's privacy and securing the exchange data difficult [49]. This system is not robust to withstand any cyberattacks, that is, the great number of stored data can be tampered, and as a result, it is not difficult to leak customer's personal information that makes the operation safety poor [50]. Moreover the payment clearing process is sophisticated, lengthy, and costly, consisting of a range of complicated procedures that cause delayed settlement with low efficiency [51]. All of the mentioned technical issues should be certainly considered, and the system should be upgraded or transformed; otherwise, these may cause notable consequences.

To address the existing challenges, the concept of blockchain is one of the most important candidates that has a potential to reshape the entire economy, considering its advantages for achieving secure exchanges between parties by **blocking** the intermediate financial institutions. The asymmetric encryption, point-to-point payment, consensus mechanisms, and the other features mentioned will make the system more secure, cost-effective, and extremely efficient [52].

Questions have been raised about how financial institutions may react to the implementations of blockchain technology into their conventional system. Surprisingly, they are not evaluated in this incorporation as a threat for the models; yet, business enterprises have attempted to modernize traditional banking system, utilizing the innovative approaches as much as possible [46]. They are seeking new avenues in this field, supporting their thoughts through the research and experimental studies for providing widespread application. For example, Rain Lohmus of Estonia's LHV bank indicated that blockchain technology is suitable for some financial implementations due to being the mostly tested and secure architecture [46]. It is also possible to point out that there are great endeavors of world's biggest banks in terms of looking for a novel and alternative system operation with blockchain and establishing a platform in financial market. The banking giants are JPMorgan, State Street, UBS, Royal Bank of Scotland, Credit Suisse, BBVA, and Commonwealth Bank of Australia, and they have an opportunity to collaborate in blockchain sector for the first time since 2015 by making considerable amount of significant contributions [53].

According to the relevant surveys including 200 global banks, it was expected that nearly 15% of banks would implement blockchain by 2017 which is extensive rate. Moreover, 66% of the banks will start to use commercial blockchain in following 4 years which is dramatic rate, has been marked by IBM, one of the most prominent companies [54].

To enhance transparency, reliability, and risk reduction, about 40 Japanese banks agreed in principle and established a consortium entitled Ripple to utilize blockchain technology in real-time transactions in a cost-effective way. The fraud events, double-spending problem, are the main issues that are avoided as possible during exchanges funds, values, or digital assets [55].

From the other perspective, blockchain-based system design should be also considered in smart grid environment for again security and data protection concerns. The state-of-the-art combination of communication and information technologies into the traditional grid makes the system "smarter" [56]. It aimed to transfer electricity in an economic and efficient fashion to commercial, social, and industrial areas and decrease the power losses in the lines. One of the most tremendous benefits of smart grid is that the system operators are always aware of the conditions of utility grid thanks to the bidirectional information flow; the smart homes, smart building, and smart campus in general smart communities have established smart devices based on AMI key technology that sends meter readings, billing, and consumption profiles to the operator.

However, there is an urgent need to eliminate cyberattack vulnerabilities of the smart grid that is suitable environment for this threat. All of the activities conducted on the Internet are evaluated as open sources of these types of attacks.

On the other hand, one of the most dramatic changes happened on the demand side, where the new actors have been integrated into the market operation, called as prosumers who are consumers also with on-site production facilities generally RESs [57]. The smart grid concept let all of the participants have communication on the online environment and pave the way for peer-to-peer energy trading between parties in a decentralized way. However, the concerns about protecting personal information are the main drawbacks for convincing the end users in terms of participating in the energy trading process.

It is highly important to ensure data transparency, data provenance, and trustworthiness between participants during peer-to-peer energy trading process, which also requires a transaction of a great amount of data in real time. Therefore blockchain technology has been found as an appropriate candidate to address the challenges. It has been indicated that this concept will help to enhance system efficiency considering current practices and procedures and also the improvement of IoT platforms can be accelerated by combining both of them in decentralized manner based on The German Energy Agency report [58].

There are considerable amount of literature studies aiming to resolve the major problems of combined blockchain and smart grid concepts. The illustration of integrating smart grid applications into blockchain is shown in Fig. 9. Tampering the meter readings is prevented by the presented studies in [59,60], which provides storage and secured the data processing. Also the studies presented by [61–62] are focused on the other main technical challenge of the system operation as high cost functions. For the purpose of increasing

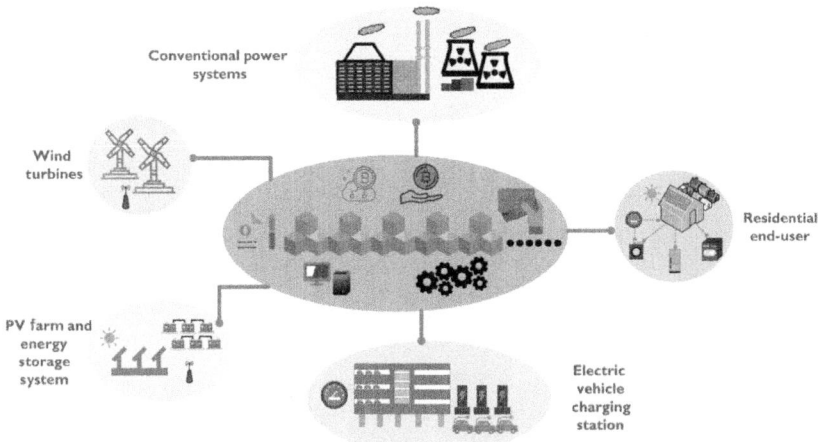

FIG. 9 The illustration of blockchain and smart grid application areas.

renewable-based energy trading in smart grid, "NRGcoin," a digital currency, was represented in [63]. The Australian company "Power Ledger" has attempted to conduct an experimental study in which peer-to-peer energy trading is executed by blockchain-based technology, which is an important example on this field [64].

Apart from mentioned implementations, blockchain technology has also gained massive importance from the academic community and utility decision-makers to operate and coordinate the system efficiently that is highly secure and cost-effective. A recently conducted survey has taken our attention to EV community and enforced the related parties to determine new business models with incorporating emergence technologies as also indicated in Table 1. Shortly the high amount of payment transactions, nearly $140 billion, has been performed for

TABLE 1 The practical initiatives for providing blockchain-based platform for EV charging [68–72].

Actor	Business	Implemented country	Generic description
MotionWerk	Start-up	Germany	Sharing the charging stations with the other EV users via mobile application, thanks to the developed Share&Charge platform
eMotorWerks	Private company	United States	Enabling peer-to-peer EV charging to the market collaborating with Share&Charge platform
Charg Coin	Tech start-up	United States	Facilitating finding the charging stations for EVs and providing secure marketplace in terms of transacting energy
Lab10 Collective	Cooperative	Austria	Fully automated charging and payment system for EVs based on blockchain technology
Easelink	Private company	Austria	Industrialize Matrix Charging with blockchain payments
Slock.it	Private company	Germany	Collaborating with MotionWerk to improve Share&Charge platform
Enexis	Utility	Netherlands	Attempting to develop IOTA-enabled transaction for performing EV charging

binding the settlements in this industry that should be taken into account [65]. A great deal of documents, data, and approvals are to be coordinated while collaboration and reallocation are ensured among vehicles with applying blockchain. Since this technology presents significant opportunities especially in security issues, the significant studies primarily aim to establish a trust-based platform and improve the level of trustworthiness [66,67].

It is evidently seen that the blockchain technology has wide range of implementation areas both financial and nonfinancial. Apart from aforementioned applications, gaming industry, notary services, cloud-based distributed storage systems, health-care services, and even in music industry are potential environments utilizing its advantages and make their system more robust, transparent, secure and resilient [46, 73].

3 Blockchain with DR applications

The electricity consumption might vary according to end-user behaviors, seasonal effects, and even a weekday/weekend difference. Independent system operators make severe efforts to respond to this variable load demand [74]. To the date of the emerging smart grid paradigm, this steadily increasing demand has tried to be tackled by considering the supply side in the traditional power system. Moreover the power system is considerably changed by including the new generation units such as PV and wind in supply side and the new players such as EVs and ESSs in demand side. Due to the new load types and the high penetration of renewable-based generation units, the imbalance between supply and demand sides must be handled to provide a more sustainable power system [75]. Therefore the operational flexibility that allows the DSO to take new axioms on both of the supply and demand sides has gained more importance in recent years, and DSM comes into prominence as a promising solution. According to the Electric Power Research Institute (EPRI), DSM can be defined as [76] "DSM is the planning, implementation and monitoring of those utility activities designed to influence customer use of electricity in ways that will produce desired changes in the utility's load shape, i.e. time pattern and magnitude of a utility's load. Utility programs falling under the umbrella of DSM include load management, new uses, strategic conservation, electrification, customer generation, and adjustments in market share."

By implementing DSM the power demand of consumers can be reduced in peak hours or shifted to the nonpeak hours as depicted in Fig. 10. To include the end users into the power system operation environment, DR strategies are accepted as the widely applied method of the DSM because of its feasibility and quick response [77]. End users are encouraged to participate in the operation of the power system in both types of incentive-based or price-based DR programs as a prosumer.

It is evidently seen that there is an unprecedented rate of incorporating DR strategies in power system operational tools to mitigate supply-demand

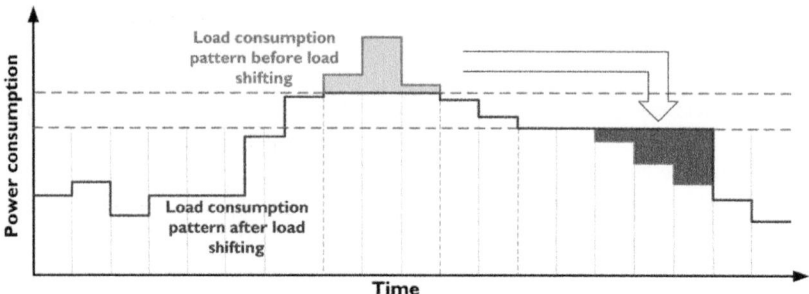

FIG. 10 The implementation of load shifting strategy by using related ancillary services.

imbalances due to unpredictable nature of RESs and power consumption patterns of end users. A considerable amount of real implementations were carried out all over the world with significant endeavors of utility decision-makers as indicated in Table 2. For more information about different DR applications, Ref. [74] can also be undertaken.

As a consequence the future electricity grid, namely, the smart grids, precisely will contain many new actors, especially in demand side to perform a more flexible operation by DR programs [78, 79]. As the number of participants increases, the system security, reliability, and efficiency together with the protection of personal data will not easily be maintained in the current version of the power system operation. At this point the significance of a blockchain-based power system will be sensed. The contract between the DSO and the DR participants can be arranged by using the blockchain so as to provide a more secure, sustainable, and reliable operation.

Numerous notable companies, stakeholders, giants of foundations, and institutions have genuinely attempted to conduct significant amount of projects and led to joint international consortiums for integrating blockchain-based DR applications into the power system. Similar to the other industries, significant changes and upheavals have been performed in both supply and demand side of the utility grid within the smart grid and blockchain concepts. The investments in RESs, DR implementations, EVs, ESSs, transactive energy models, and other application domains have been increased in recent years, coming from energy transition needs from traditional to the modernized grid architecture.

To harness the large number and variety of flexibility sources demand reduction capability in a more organized and coordinated way, TenneT (a transmission system operator in Germany), Vandebron, Sonnen, and IBM have joint a consortium that started in 2017. The pilot projects were implemented in the real world and tested in the Netherlands and Germany for the purpose of increasing system performance and achieving system in balance by utilizing the flexibility of electric cars and home batteries. Peer-to-peer trading platforms, open-source for blockchain provider, and batteries are provided by stakeholders jointly involved in the project [64].

TABLE 2 DR implementations in practical scale.

Focused on controlling a specific loads	Air conditioners	Pacific Gas & Electric Company [PG&E (California)], Commercial & Residential—"*SmartAC program*"
		CPS Energy (Texas), Commercial & Residential, "*Smart Thermostat Program*"
		Austin Energy Company (Texas), "Rush Hour Rewards Program"
		Energex Company (Australia), "PeakSmart AC program"
	Pool pumps	Endeavour Energy (Australia), Residential, "PoolSaver Program"
		Energex Company (Australia), Residential, "Pool Rewards Program"
	Water heaters	Energex Company (Australia), Residential, "Hot Water Rewards Program"
	Irrigation pumps	Southern California Edison (SCE) Company, "Agricultural and Pumping Interruptible Program"
		Transpower Company (New Zealand)
Focused on controlling the total power consumption of end-user premises	Contract-based reduction	PG&E (California)—Industrial, "Optional Binding Mandatory Curtailment Program"
		Diamond Energy Company (Singapore), "Load Interruption Program"
		Ausgrid Company (Australia), Commercial & Industrial, "Dynamic Peak Rebate Trial"
	Building energy management system	Kyocera, IBM Japan & Tokyo Community (Japan), Converge & OpenADR Alliance & Fujitsu, pilot implementation
		Southern California Edison (SCE) Company (California), Commercial, "Automated DR Program"
	Backup generators	TECO & Progress Energy Company, (Florida), "Backup Generator Program"
		Eskom Company (South Africa), "Standby Generator Program"

Continued

TABLE 2 DR implementations in practical scale — cont'd

Focused on enabling end users to the electricity market directly or indirectly	New York Independent System Operator (NYISO), (New York), "Day-Ahead DR Program"
	New York Independent System Operator (NYISO)—(New York), "Demand Side Ancillary Services Program"
	PG&E (California), Industrial, "Demand Bidding Program"
	Southern California Edison (SCE) Company, (California), Commercial, "Demand Bidding Program"

The blockchain concept has been evaluated as an attractive solution in modern energy market and an important step comes from the Energy Web Foundation ("EWF"), one of the most widely known foundations in the world that creates Energy Web Chain platform for promoting the EVs, renewable energy credits, DR applications, distributed generation system, and the like in the energy industry [80]. Belgium's TSO, namely, Elia, is also an affiliate of EWF and has been aware of the challenges in emerged sophisticated power system in terms of ensuring production and consumption balance while introducing thousands and eventually millions of assets and variable RES installations. As a solution the concept of demand-side flexibility has been taken into consideration and also the pilot project was carried out for commercial and industrial end users in 2013 by Elia. Recently, there is a significant endeavor to construct a blockchain-based application for implementing DR programs automatically considering the grid requirements in balancing market. The developed architecture is aimed to be performed on Tobalaba, EWF's blockchain test network. Sam Hartnett, a member of the EWF team and an associate at Rocky Mountain Institute, indicated his thoughts as follows: "When a grid operator like Elia introduces a new technology like blockchain and puts a strong foot forward, the whole industry adapts-service providers, aggregators, consumers. They've put a flag in the ground for blockchain's potential to change the market." [81].

Spectral Smart Energy Control Systems have been presented as such an advanced energy management platform that is capable of controlling wide range of energy assets available in the smart grid from distributed local energy resources, battery systems, and smart home appliances to even aggregated mobile loads. One of the most important targets of the project is establishing a blockchain-based market platform for maximizing RES penetration and flexibility resources effectiveness by empowering individual prosumers and/or emerged local energy communities to being an active participant in energy market. The intermediate parties are eliminated, which bring direct, automated, and robust peer-to-peer trading mechanism with complete transparency in their identity and trading details thanks to the developed Spectral Energy Xchange platform [82].

The project of eDREAM has been qualified to be supported by European Union's Horizon 2020 research and innovation program with the aim of developing a novel near-real-time closed loop optimal blockchain-based DR ecosystem for aiding distribution system operators (DSOs) to operate system under reliability requirements and to maintain its secure, sustainable conditions. The considerable amount of possible flexibility resources has been utilized in ancillary and balancing services while ensuring optimal system operation by aggregators. The blockchain applications have been investigated in decentralized marketplace-driven management with securing the data handling. For testing the developed architecture, three pilot areas are determined in the United Kingdom, Italy, and Greece corporating the Kiwi, Terni, and CERTH Lab Facilities, respectively. For more detailed information, Ref. [83] can be examined.

The electricity market architecture has been transfigured from conventional model to novel decentralized and autonomous energy sector that has quite differences with from each other. To address the need of creating a blockchain-powered flexible peer-to-peer trading platform, a London-based start-up Electron launched its adventure in the year 2015 in cooperation with National Grid UK and Flexitricity. For increasing the deployment smart grid applications into the power system, they also present some products that are capable of registrating the meters, trading DR event actions, and managing the distributed energy resources—making all of them easy to trade between distributed parties in an efficient manner. The consortium targets to take the advantage of blockchain technology as an enabler of flexibility while ensuring security and transparency of the developed decentralized structure [84]. Alastair Martin, a member of Flexitricity, has indicated as follows: "One of the issues faced by the U.K. energy sector today is metering which measures the contribution of demand-response and could be potentially contaminated by unrelated factors. Submetering is key and would allow us to determine the correct level of energy delivery at site, but this requires appropriate information flow and validation. Blockchain technology has the potential able to address this issue successfully and enable us to fully optimise demand-response" [68].

It can be deduced from the aforementioned explanations that DR program implementations have some points needed to be improved such as the energy intermediaries (aggregators or DSOs) that provide communication with end users on an individual basis. However, this might cause the success rate of the DR to decrease; that is, the received demand reduction requests coming from electric utilities cannot be met by end users for achieving their power-saving targets. Being aware of the mentioned issues, Japanese companies Fujitsu Limited and Fujitsu Laboratories Ltd. have made an effort to devise a blockchain-based system for enabling enterprise end users to perform peer-to-peer energy exchange among each other. The developed architecture has applied the system based on real-world electricity data supported also be ENERES Co., Ltd., and as a result, DR success rate was nearly improved by 40% compared with existing implementation. It is expected that the considerable improvement will pave the way for increasing the number of participants into the DR programs that help to give fast responses to the changes especially in peak periods [85].

Ethereum and smart contract-based transActive Grid (TAG) platform was developed by the LO3 Energy, which makes neighboring residents to perform localized peer-to-peer energy trading and control its distributed resources in grid balancing axioms by implementing DR applications on Brooklyn Microgrid (one of the most famous pilot projects developed by collaboration of LO3 energy and Siemens). To monitor and measure the prosumers' energy variations, transmitting this information to the other end-user nodes in created network and acting upon the available information, TAG elements were designed including computer and meter. The excessive available solar energy produced by end users can be directly transacted to theirs neighbors for improving system performance beginning from the local level [86].

Optimal operation requirements of distribution power system are taken into account thoroughly, and a platform is aimed to be created by the Hive Power to develop optimal management strategies considering the revolutions, changes, and upheavals on system the model. The participants are incentivized to contribute maintaining the electricity system in balance by effective coordination of their production and consumption values. One of the important features of this developed Ethereum blockchain is that flexibility sources of the local energy communities have been managed optimally by satisfying supply-demand needs for the purpose of maximizing entire community's welfare and meeting system technical constraints thanks to the HONEY algorithm. Also, Hive Token is used in energy trading activities, which is standard Ethereum ERC20 token [87].

The sophisticated architecture of energy sector has different necessities that should be considered from general perspectives such as existence of vast range of data to be processed fast and need a considerable amount of storage capacity and the like. To address these specific challenges, again, Ethereum-based blockchain was designed by the Pylon Network especially for assisting energy suppliers to get better information about power flows and also provide renewable energy cooperatives. Similar to Hive Power platform, the scalable and versatile system also promotes smart meter (Metron) in combination with blockchain and virtual trades, energy flows are tokenized. Surely, it is possible to dispatch the demand for achieving optimal power flows in real time in distributed green energy network [88].

4 Concluding remarks

It can be evidently seen that 21st century brings along significant revolutions in nearly every industry especially power system, which all are forerunner of modern society from top to bottom. The long-lasting structures have been transfigurated extremely different ones, making it necessary to introduce brand-new technologies for keeping up with rapidly changing time. Highly penetrated RESs, demand-side management programs, EV integration with both vehicle-to-grid and grid-to-vehicle options, peer-to-peer trading platforms, IoT-enabled smart appliances, flexible control and management systems, machine learning algorithms and others have enforced energy sector stakeholders to find advanced solutions for achieving operate the system in desired targets. The current power system and financial and nonfinancial structures have been transformed from centralized to decentralized and distributed network platforms that also enable information flows at significant number of terminals. Therefore, as the trending topology, the usage areas of blockchain are being widened in nowadays because of the changing architectures, making it necessary to combine advanced technological innovations for managing them in a secure, transparent, versatile, scalable, and authenticated fashion. In this context, blockchain has become one of the latest "disruptive innovations," and it has taken great attention from utility decision-makers, financial institutions, national governments, the academic community, and industrial stakeholders due to its high potential in terms of reconfigurating society entirely.

From the other perspective, DR programs are being used to decrease the electric power consumption in peak periods or shift the power demands to the nonpeak periods of the day since its invention. The number of participants in such programs will increase in the future due to the increment of the self-energy production or the desire to control their consumptions. Thus the lack of security in that system including many stakeholders during the operation of the power system is inevitable. In this regard, the incorporated both the blockchain and DR programs can be accepted as a promising solution so as to provide a more secure and sustainable power transaction.

Overall, there have been great attempts to map out the early stages of power system, drawbacks, changing features, incorporated new technological innovations, and the background of shifting distributed structures, comprehensively. Also the fundamental features of blockchain technology were holistically examined including consensus algorithms, asymmetric cryptography techniques, specific hash functions, and peer-to-peer transaction network. The high diffusion within the business, financial, nonfinancial fields was evaluated from different points of view. Moreover the real practical implications, projects, and startups were elaborated in detail for emphasizing whether the blockchain is only a buzzword or has a chance in a community especially combined with DR strategies.

Acknowledgments

The authors would like to convey special thanks for creative icons developed by anonymous creators with nicknames Turkkub, flat ıcons, payungkead, smashicons, freepik, payungkead, prettycons, Icongeek26, monkik, srip, and eucalyp. J.P.S. Catalão acknowledges the support by the FEDER funds through COMPETE 2020 and by the Portuguese funds through FCT, under POCI-01-0145-FEDER-029803 (02/SAICT/2017).

References

[1] B.E. Santacana, G. Rackliffe, X. Feng, Getting smart with a clearer vision of the intelligent grid, control emerges from chaos, IEEE Power Energy Mag. 8 (2010) 41–48.

[2] R. Ma, H. Chen, Y. Huang, W. Meng, S. Member, Smart grid communication: its challenges and opportunities, IEEE Trans. Smart Grid 4 (2013) 36–46.

[3] Q. Li, The future-oriented grid-smart grid, J. Comput. 6 (2011) 98–105.

[4] A.K. Erenoğlu, O. Erdinç, A. Taşcıkaraoğlu, Pathways to a Smarter Power System: History of Electricity, 1-45 Elsevier Science, Oxford/Amsterdam, 2019, p. 2019.

[5] J. Arias-gaviria, S.X. Carvajal-quintero, S. Arango-aramburo, Understanding dynamics and policy for renewable energy diffusion in Colombia, Renew. Energy 139 (2019) 1111–1119.

[6] M. Guidolin, T. Alpcan, Transition to sustainable energy generation in Australia: interplay between coal, gas and renewables, Renew. Energy 139 (2019) 359–367.

[7] K. Hansen, C. Breyer, H. Lund, Status and perspectives on 100% renewable energy systems, Energy 175 (2019) 471–480.

[8] G. Tuna, V.E. Tuna, The asymmetric causal relationship between renewable and non-renewable energy consumption and economic growth in the ASEAN-5 countries, Resour. Policy 62 (2019) 114–124.

[9] A.S. Dagoumas, N.E. Koltsaklis, Review of models for integrating renewable energy in the generation expansion planning, Appl. Energy 242 (2019) 1573–1587.

[10] S. Jenniches, Assessing the regional economic impacts of renewable energy sources—a literature review, Renew. Sustain. Energy Rev. 93 (2018) 35–51.

[11] A. Eitan, L. Herman, Community–private sector partnerships in renewable energy, Renew. Sustain. Energy Rev. 105 (2019) 95–104.

[12] S. Yin, J. Wang, F. Qiu, Decentralized electricity market with transactive energy—a path forward, Electr. J. 32 (2019) 7–13.

[13] S. Guo, Q. Liu, J. Sun, H. Jin, A review on the utilization of hybrid renewable energy, Renew. Sustain. Energy Rev. 91 (2018) 1121–1147.

[14] B. Yu, J. Guo, C. Zhou, Z. Gan, J. Yu, F. Lu, A review on microgrid technology with distributed energy, in: Proc.—2017 Int. Conf. Smart Grid Electr. Autom. ICSGEA, 2017, pp. 143–146.

[15] A. Ravichandran, P. Malysz, S. Sirouspour, A. Emadi, The critical role of microgrids in transition to a smarter grid: a technical review, in: IEEE Transp. Electrif. Conf. Expo Components, Syst. Power Electron.—From Technol. to Bus. Public Policy, ITEC, 2013.

[16] M. Faisal, M.A. Hannan, P.J. Ker, A. Hussain, M. Bin Mansor, F. Blaabjerg, Review of energy storage system technologies in microgrid applications: issues and challenges, IEEE Access 6 (2018) 35143–35164.

[17] V. Lakshminarayanan, V.G.S. Chemudupati, S. Pramanick, K. Rajashekara, Real-time optimal energy management controller for electric vehicle integration in workplace microgrid, IEEE Trans. Transp. Electrif. (2018) 1-1.

[18] S.A. Pourmousavi, M.H. Nehrir, Demand response for smart microgrid: initial results, in: IEEE PES Innov. Smart Grid Technol. Conf. Eur. ISGT Europe, 2011, pp. 11–16.

[19] K.T.M.U. Hemapala, A. Kulasekera, Demand side management for microgrids through smart meters, in: Proceedings of the IASTED Asian Conference on Power and Energy Systems, AsiaPES 2012, 2012. https://doi.org/10.2316/P.2012.768-060.

[20] H.J. Cha, J.Y. Choi, D.J. Won, Smart load management in demand response using microgrid EMS, in: ENERGYCON 2014—IEEE Int. Energy Conference, 2014, pp. 833–837.

[21] M. Fan, X. Zhang, Consortium blockchain based data aggregation and regulation mechanism for smart grid, IEEE Access 7 (2019) 35929–35940.

[22] M.M. Esfahani, S. Member, O.A. Mohammed, Secure blockchain-based energy transaction framework in smart power systems, in: IECON 2018—44th Annual Conference of the IEEE Industrial Electronics Society, vol. 1, IEEE, 2018, pp. 260–264.

[23] A.S. Musleh, S.M. Muyeen, Blockchain applications in smart grid—review and frameworks, IEEE Access 7 (2019) 86746–86757.

[24] M. Andoni, V. Robu, D. Flynn, S. Abram, D. Geach, D. Jenkins, P. Mccallum, A. Peacock, Blockchain technology in the energy sector: a systematic review of challenges and opportunities, Renew. Sustain. Energy Rev. 100 (2019) 143–174.

[25] L. Yang, The blockchain: State-of-the-art and research challenges, J. Ind. Inf. Integr. 15 (2019) 80–90.

[26] F. Glaser, L. Bezzenberger, Beyond cryptocurrencies—a taxonomy of decentralized consensus systems, in: 23rd European Conference on Information Systems, 2015, pp. 1–18.

[27] S. Muftic, I. Sanchez, L. Beslay, Overview and Analysis of the Concept and Applications of Virtual Currencies, 2016, (JRC Technical Report).

[28] UK Government, Government Office of Science, Distributed Ledger Technology: Beyond Blockchain, 2016 (Report).

[29] J.A. Castellanos, D. Coll-mayor, J.A. Notholt, Cryptocurrency as guarantees of origin: simulating a green certificate market with the Ethereum Blockchain, (2017). https://doi.org/10.1109/sege.2017.8052827.

[30] W. Diffie, M. Hellman, New directions in cryptography, IEEE Trans. Inf. Theory 22 (6) (1976) 644–654.

[31] Bitcoin Public and Private Keys, (Online), Available from: https://www.dummies.com/software/other-software/bitcoin-public-private-keys/, 2019. (Accessed October 2019).

[32] K. Christidis, G.S. Member, Blockchains and smart contracts for the Internet of Things, IEEE Access 4 (2016) 2292–2303.

[33] Blockchain Hash Function, (Online), Available from: https://www.javatpoint.com/blockchain-hash-function, 2019. (Accessed October 2019).

[34] How Cryptographic Algorithms and Hashing Keep Blockchain Secure, (Online), Available from: https://jaxenter.com/cryptographic-hashing-secure-blockchain-149464.html, 2019. (Accessed October 2019).

[35] Blockchain: How Mining Works and Transactions are Processed in Seven Steps, (Online), Available from: https://blog.goodaudience.com/how-a-miner-adds-transactions-to-the-blockchain-in-seven-steps-856053271476, 2019. (Accessed October 2019).

[36] M. Walport, Distributed Ledger Technology: Beyond Blockchain, (Online), Available from: https://assets.publishing.service.gov.uk/government/uploads/system/uploads/attachment_data/file/492972/gs-16-1-distributed-ledger-technology.pdf, 2019. (Accessed October 2019).

[37] T.U. Tulsıdas, Smart Contracts From a Legal Perspective, (Online), Available from: https://pdfs.semanticscholar.org/370b/bc9f7af31fa01a8f61276d2a67591d22680e.pdf, 2019. (Accessed October 2019).

[38] Y. Hou, Y. Chen, Y. Jiao, J. Zhao, H. Ouyang, P. Zhu, D. Wang, Y. Liu, A resolution of sharing private charging piles based on smart contract, in: 2017 13th International Conference on Natural Computation, Fuzzy Systems and Knowledge Discovery (ICNC-FSKD), IEEE, 2017, pp. 3004–3008.

[39] Smart Contracts—A Time Saving Primer, (Online), Available from: https://hackernoon.com/smart-contracts-a-time-saving-primer-b3060e3e5667, 2019. (Accessed October 2019).

[40] A Quick Guide to Understanding Blockchain Smart Contracts, (Online), Available from: https://thenextweb.com/hardforkbasics/2019/03/22/a-quick-guide-to-understanding-blockchain-smart-contracts/, 2019. (Accessed October 2019).

[41] Blockchain Smart Contracts Aren't Smart and Aren't Contracts, (Online), Available from: https://www.forbes.com/sites/davidblack/2019/02/04/blockchain-smart-contracts-arent-smart-and-arent-contracts/#b1d217c1e6a8, 2019. (Accessed October 2019).

[42] Blockchainhub Berlin, Smart Contracts, (Online), Available from: https://blockchainhub.net/smart-contracts/, 2019. (Accessed October 2019).

[43] How Bitcoin Grew Up and Became Big Money, (Online), Available from: https://www.theverge.com/platform/amp/2019/1/3/18166096/bitcoin-blockchain-code-currency-money-genesis-block-silk-road-mt-gox, 2019. (Accessed October 2019).

[44] S. Nakamoto, Bitcoin: A Peer-to-Peer Electronic Cash System, (Online), Available from: https://bitcoin.org/bitcoin.pdf, 2008. (Accessed October 2019).

[45] M. Nofer, P. Gomber, O. Hinz, D. Schiereck, Blockchain—a disruptive technology, Bus. Inf. Syst. Eng. 59 (2017) 183–187.

[46] M. Crosby, P. Pattanayak, S. Verma, V. Kalyanaraman, BlockChain technology: beyond Bitcoin, Appl. Innov. Rev. (2016) 5–20.

[47] T. Aste, P. Tasca, T. Di Matteo, Blockchain technologies: the foreseeable impact on society and industry, Computer 50 (2017) 18–28.

[48] S. Aggarwal, R. Chaudhary, G.S. Aujla, N. Kumar, K.K.R. Choo, A.Y. Zomaya, Blockchain for smart communities: applications, challenges fnd opportunities, J. Netw. Comput. Appl. 144 (2019) 13–48.

[49] D. Bradbury, Blockchain's big deal, Eng. Technol. 11 (10) (2016).

[50] Y. Guo, C. Liang, Blockchain application and outlook in the banking industry. Financ. Innov. (2016), https://doi.org/10.1186/s40854-016-0034-9.

[51] Ernst & Young, Overview of Blockchain for Energy and Commodity Trading, http://www.ey.com/Publication/vwLUAssets/ey-overview-of-blockchain-forenergy-and-commodity-trading/FILE/ey-overview-of-blockchain-for-energy-and-commodity-trading.pdf, 2017. Accessed 20 November 2017. (Accessed October 2019).

[52] Q.-G. Mu, Second Report on Survey of Blockchain Technology: Evolution of Blockchain Technology, Report by Chuancai Securities Co., Ltd., 2016.

[53] J. Kelly, Nine of World's Biggest Banks Join to Form Blockchain Partnership, Reuters, Thomson Reuters, 2016.

[54] Fortune: Blockchain Will Be Used by 15% of Big Banks By 2017, http://fortune.com/2016/09/28/blockchain-banks-2017/, 2019. 2016-09-28/2016-10-23 (Accessed October 2019).

[55] Q. Wang, X. Zhu, Y. Ni, L. Gu, H. Zhu, Blockchain for the IoT and Industrial IoT: A Review, Internet of Things, 2019.

[56] M.-L. Marsal-Llacuna, Future living framework: is blockchain the next enabling network? Technol. Forecast Soc. Change 128 (2018) 226–234.

[57] EU Commission, Study on Residential Prosumers in the European Energy Union, https://ec.europa.eu/commission/sites/beta-political/files/study-residentialprosumers-energy-union_en.pdf, 2017. (Accessed April 2019).

[58] C. Burger, A. Kuhlmann, P. Richard, J. Weinmann, Blockchain in the Energy Transition a Survey Among Decision-Makers in the German Energy Industry, https://shop.dena.de/fileadmin/denashop/media/Downloads_Dateien/esd/9165_, 2016. (Accessed October 2019).

[59] J. Gao, K.O. Asamoah, E.B. Sifah, A. Smahi, Q. Xia, X. Zhang, Grid monitoring: secured sovereign blockchain based monitoring on smart grid, IEEE Access 6 (2018) 9917–9925.

[60] S. Aggarwal, R. Chaudhary, G.S. Aujla, A. Jindal, A. Dua, N. Kumar, Energy chain: enabling energy trading for smart homes using blockchain. (2018). https://doi.org/10.1145/3214701.3214704.

[61] M. Mylrea, S.N.G. Gourisetti, Blockchain for smart grid resilience: exchanging distributed energy at speed, scale and security, (2017). https://doi.org/10.1109/RWEEK.2017.8088642.

[62] F. Lombardi, L. Aniello, S. De Angelis, A. Margher, V. Sassone, A blockchain based infsrastructure for reliable and cost-effective iot-aided smart grids, (2018). https://doi.org/10.1049/cp.2018.0042.

[63] Z. Li, J. Kang, R. Yu, D. Ye, Q. Deng, Y. Zhang, Consortium blockchain for secure energy trading in industrial Internet of Things, IEEE Trans. Ind. Inf. 14 (2018) 3690–3700.

[64] L. Diestelmeier, Changing power: shifting the role of electricity consumers with blockchain technology—policy implications for EU electricity law, Energy Policy 128 (2019) 189–196.

[65] A.M. Wyglinsk, J. Irvine, J. Chapman, The future of vehicular security and privacy, IEEE Veh. Technol. Mag. 13 (2018).

[66] S.K. Datta, Vehicles connected resources: opportunities and challenges for the future, IEEE Veh. Technol. Mag. 12 (2) (2017) 26–35.

[67] L. Hong, Z. Yan, Y. Tao, Blockchain-enabled security in electric vehicles cloud and edge computing, IEEE Netw. 32 (3) (2018) 78–83.

[68] World Energy Insights Brief, World Energy Council, https://www.worldenergy.org/assets/downloads/World-Energy-Insights-Blockchain-Anthology-of-Interviews.pdf, 2018. (Accessed January 2020).

[69] P2P Vehicle Charging: Is Blockchain a Driver of EV Adoption? https://powertechresearch.com/p2p-vehicle-charging-is-blockchain-a-driver-of-ev-adoption/, 2018. (Accessed January 2020).

[70] Energy is Money, https://chgcoin.org/, 2020. (Accessed January 2020).

[71] About the lab10 Collective, https://lab10.coop/about/about-us/, 2020. (Accessed January 2020).

[72] Blockchain Innovation Landscape Brief, International Renewable Energy Agency, https://www.irena.org/-/media/Files/IRENA/Agency/Publication/2019/Feb/IRENA_Landscape_Blockchain_2019.pdf?la=en&hash=1BBD2B93837B2B7BF0BAF7A14213B110D457B392, 2020. (Accessed January 2020).

[73] T. McGhin, K.K.R. Choo, C.Z. Liu, D. He, Blockchain in healthcare applications: research challenges and opportunities, J. Netw. Comput. Appl. 135 (2019) 62–75.

[74] N.G. Paterakis, O. Erdinç, J.P. Catalão, An overview of demand response: key-elements and international experience, Renew. Sustain. Energy Rev. 69 (2017) 871–891.

[75] P. Du, N. Lu, H. Zhong, Demand Response in Smart Grids, Springer, 2019.

[76] Electric Power research Institute (EPRI)—Definition of Demand-Side Management, http://www.epri.com, 2019. (Accessed September 2019).

[77] A. Losi, P. Mancarella, A. Vicino, Integration of Demand Response into the Electricity Chain: Challenges, Opportunities, and Smart Grid Solutions, John Wiley & Sons, 2015.

[78] X. Yang, G. Wang, H. He, J. Lu, Y. Zhang, Automated demand response framework in ELNs: decentralized scheduling and smart contract. IEEE Trans. Syst. Man Cybern. Syst. (2019) 58–72, https://doi.org/10.1109/TSMC.2019.2903485.

[79] Z. Zhou, B. Wang, Y. Guo, Y. Zhang, Blockchain and computational intelligence inspired incentive-compatible demand response in internet of electric vehicles, IEEE Trans. Emerg. Top. Comput. Intell. 3 (3) (2019) 205–216.

[80] Blockchain Continues to Make Headway in the Energy Industry, (Online), Available from: https://www.lawoftheledger.com/2019/07/articles/blockchain/blockchain-technology-energy-industry/, 2020. (Accessed January 2020).

[81] Belgium's Transmission System Operator Eyes Blockchain for Demand Response, (Online), Available from: https://energyweb.org/2018/02/28/belgiums-transmission-system-operator-eyes-blockchain-for-demand-response/, 2020. (Accessed January 2020).

[82] Spectral Smart Building Platform, (Online), Available from:www.spectral.energy, 2020. (Accessed January 2020).

[83] Online, Available from: https://edream-h2020.eu/demand-response-tools/ (Accessed January 2020).

[84] Energy Industry and Blockchain: Overview and Applications, (Online), Available from: https://hackernoon.com/energy-industry-and-blockchain-overview-and-applications-c9c88 e2039b, 2020. (Accessed January 2020).

[85] Fujitsu Develops Blockchain-Based Exchange System for Electricity Consumers, (Online), Available from: https://www.fujitsu.com/global/about/resources/news/press-releases/2019/ 0130-01.html, 2020. (Accessed January 2020).

[86] A. Goranovic, M. Meisel, A. Treytl, T. Sauter, Blockchain applications in microgrids: an overview of current projects and concepts, in: IECON 2017—43rd Annual Conference of the IEEE Industrial Electronics Society, 2017. https://doi.org/10.1109/IECON.2017.8217069.

[87] M. Pichler, M. Meisel, A. Goranovic, K. Leonhartsberger, G. Lettner, G. Chasparis, H. Vallant, S. Marksteiner, H. Bieser, Decentralized Energy Networks Based on Blockchain: Background, Overview and Concept Discussion, Lecture Notes in Business Information Processing, vol. 339, Springer, Cham, 2019.

[88] Pylon Network Whitepaper, (Online), Available from: https://pylon-network.org/wp-content/ uploads/2017/07/170730_WP-PYLON_EN.pdf, 2020. (Accessed January 2020).

Chapter 7

Blockchain-based demand response using prosumer scheduling

Hosna Khajeh, Miadreza Shafie-khah and Hannu Laaksonen
School of Technology and Innovations, University of Vaasa, Vaasa, Finland

1. Introduction

New technological advances make significant changes in communication between customers and different stakeholders in the power system. These developments pave the way for the power system, heading toward the new age of smart digital grids. Smart meters connected to customer's appliances enable customers to manage their consumption to maximize the profit. In addition, customers can also play an active role in local markets, submit bids, and reshape their demand based on certain signals such as market prices. Demand response (DR) programs help customers to manage their loads in a way that they can maximize their profit. The customers not only are able to control their consumption but also play the role of producer in future smart grids. For example, a household can be equipped with solar panels so as to supply its demand and sell the extra power to the grid. In this way, customers change to "prosumers" considered as actors who can participate in the market as a seller or a buyer.

More proactive role of customers can bring considerable benefits to the power system. First, it reduces the need of installing new generation capacity for situations where total demand exceeds the generation. In the future, building of new large conventional fossil fuel-based power plants will not be economically viable and environmentally acceptable solution [1–4]. From environmental reasons, large-scale installation of renewables is needed, but the intermittent characteristics of renewables need new solutions based on, for example, short- and long-term energy storages and on demand response utilization. Intelligent control and management of flexible demand as one important part of the solution would save the total power system costs and enable affordable energy prices for the customers [5]. Second, participating in a DR program may mitigate the price spikes during high demand, i.e., peak load hours [6, 7].

Blockchain-based Smart Grids. https://doi.org/10.1016/B978-0-12-817862-1.00007-5

Through more active role of prosumers in the future power systems and electricity markets, the power system resiliency and demand elasticity can be increased, and more competitive customer electricity prices can be achieved.

However, the customers prefer to operate their appliances in a way that ensures that the highest comfort is reached [8]. The main responsibility of a DR program is to motivate customers to respond to the market prices despite the fact that it may reduce their comfort, helping the operator keep the balance between the generation (or supply) and demand [9, 10]. The current DR scheme is classified into price-based and incentive-based programs. In the incentive-based program, a contractual agreement may be reached between two parties, or the utility company may request for reshaping the demand [11]. On the contrary the utility company has an indirect role in the price-based DR program. The customers try to adjust their consumption according to the prices determined by the utility [12]. The main reasons encouraging customers to take part in a DR program are saving costs or making profits, avoiding blackout (power system resiliency), and responsibility sensing [13].

Implications of both price-based and incentive-based DR programs have been thoroughly assessed in several studies. For instance, in [14], real-time pricing was utilized to incentivize customers at peak load hours. The research shows that peak load could decrease between 8 and 11 h in the UK residential sector. In [15] a market deploying real-time pricing was modeled, and a reduction in energy costs could be achieved if a considerable number of participants respond to the market prices. The real-time pricing tariff was proposed to be set by the utility in [16]. The study aims to maximize the utility's profit taking into account the price elasticity of the customers. Different DR programs were offered to be prioritized considering the nodal and global factors in [17]. In [18], responsive loads offer prices and quantities of the demand so as to be curtailed during peak load time slots or be shifted to another hour. An incentive-based DR algorithm was introduced by [19], in which the incentive rates are paid to the customers depending on the amount of the shifted load and improved voltage profile. In [20] the electricity retailer tries to manage the market price risks through incentive-based DR program. The rationality and incentive compatibility of a single customer with an incentive-based DR were assessed in [21]. The Stackelberg game was adopted by Refs. [22, 23] so as to model the interaction between different agents (or stakeholders/actors), like the grid operator and customers, from the perspective of the grid operator utilizing an incentive-based DR program.

Since the direct negotiation between the system operator and small customers or prosumers would make the system more complex, a linking market or agent can be in charge of aggregating the customers' responsive loads. In a centralized approach a prosumer is not able to participate directly in the DR program. It should share the information with the centralized entity (e.g., an aggregator) so that the entity can decide how to manage the prosumer's consumption taking into account the information given by the prosumer.

The majority of studies conducting research on DR programs have considered the centralized approach. In a decentralized DR scheme, however, a customer or prosumer can directly have control over its demand. In a decentralized approach, local markets are constituted in a way that it facilitates the trading between small customers and prosumers. The direct participation of them can increase liquidity, promoting collaborative trading in the market. In addition, a customer or prosumer will have more freedom to innovate and choose the level of discomfort cost it can incur. The decentralized program may also eliminate conflicts between the prosumers and the upstream entity and increase electricity supply security and reliability as well as power system resiliency.

In the literature, there exist some studies that propose a decentralized scheme for DR deployment. In [24] a decentralized DR framework is proposed considering each bus as an individual agent. However, customers are not capable to react individually, and therefore a bus service entity was proposed to submit bids on behalf of the customers connected to the bus. Similar kind of decentralized approach-based study regarding the renewable generation and responsive demand management has been done in [25]. Again, load aggregators are responsible for aggregating small customers' responsive loads.

The main contribution of this chapter is as follows:

1. Each prosumer with responsive demand and renewables is considered as an individual agent that will be able to submit bids in the real-time local market.
2. DR is implemented in a decentralized scheme in which customers can totally control their consumption, reshaping their demand at their convenience according to the market prices. The type of DR program proposed in this chapter is price based.
3. A Q-learning method is utilized by the prosumer so as to learn how to build its bids.
4. Prosumers can compete against each other aiming to maximize their own profits in the market that increase the liquidity of the local market.
5. A blockchain-based platform is adopted to facilitate the participation of prosumers, promoting trust and privacy among the members of the market while eliminating a broker (like aggregator).
6. The proposed decentralized model is compared with the centralized aggregated-based one that is considered as an incentive-based DR program.

The rest of this chapter is organized in the following way: Section 2 introduces the proposed model briefly. Section 3 describes the first stage in which prosumers are scheduled through the Q-learning method, and the market is settled to find the best market price and DR price. Section 4 explains the second stage of the proposed model where demand is rescheduled and the market is settled for the last time. Section 5 gives details about how the proposed model can be implemented on the blockchain platform. Section 6 compares the proposed decentralized model with the aggregator-based scheme and assesses the effect of prosumers' contribution on the market's load profile. Section 7 concludes the chapter.

2. The decentralized demand response model

In this model, prosumers are assumed to take part in a local privacy-based market. Accordingly the main model is from the viewpoint of the prosumers. However, the local market-clearing model is also presented in this work. Prosumers of the same market can trade with each other in a blockchain-based platform. There exists no intermediary intervening in the trading. One of the key features of blockchain technology, the smart contract, is adopted to implement the settlement rules related to the local market. Fig. 1 illustrates the model and the interaction between prosumers in the local market.

Our proposed model includes two stages:

- In the first stage a prosumer tries to find the optimal price and capacity for the next hour through the use of the Q-learning technique. It should be noted that the prosumers are proposed to utilize the maximum capacities of their renewables since they are more profitable due to their low marginal costs and environmental benefits [26]. Then, prosumers submit their bids via the blockchain account. Eventually the market is settled through the use of the smart contracts, and the next-hour market price and the DR price are sent to the participants.
- In the second stage, prosumers are given the option to reshape their demand based on the market and DR price, aiming to maximize their profit with regard to their own information. Again the market will be settled trying to maximize the social welfare of the participants.

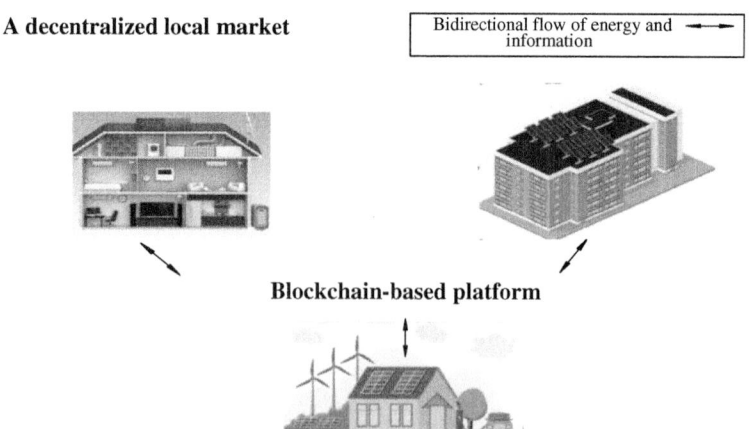

FIG. 1 The proposed decentralized local market containing various prosumers.

3. Stage 1: The prosumer's scheduling and the first settlement

The prosumers' scheduling will be performed through the reinforcement learning technique. It also calls deep Q-learning with experience replay. With regard to this technique, a prosumer would be able to find its optimal bidding strategy through the use of experiences it has gained previously. In other words the previous interactions of the participant with the market help to set the optimum bidding strategy. All of the data associated with the past experiences including the status of the past and the current ones as well as the actions and the related rewards are stored in a list called experience replay. During each episode, one part of the list is deployed randomly so as to train the prosumer, helping the algorithm avoid overfitting. The final purpose of the algorithm is to predict the Q-function, leading to find future reward. Fig. 2 shows the pseudocodes employed to predict the future reward through finding the optimum Q-function [27]. Note that the Q-values are the estimated reward, taken to be independent of the system's states to reduce the complexity of the proposed model.

Algorithm Deep Q-learning with experience replay

Put the initial value for replay memory D

Put the initial value for action-value function Q with random factors

for episode = 1, M
- Initialize sequence $s_1 = \{x_1\}$
- Initialize preprocessed sequence $\phi_1 = \phi(s_1)$

   for $t = 1, T$
- Select a random action a_t with the probability ρ
- Otherwise $a_t = \max_a Q^*(\phi(s_t), a; \theta)$ with the probability $1 - \rho$
- Implement a_t and observe reward r_t and image x_{t+1}
- Set $s_{t+1} = s_t, a_t, x_{t+1}$ and preprocess $\phi_{t+1} = \phi(s_{t+1})$
- Store transition $(a_t, r_t, \phi_t, \phi_{t+1})$ in D
- Sample random part of transitions $(a_i, r_i, \phi_i, \phi_{i+1})$ from D
- Set $y_i =$
$$\begin{cases} r_i & \text{For terminal } \phi_{i+1} \\ r_i + \pi_t maxQ(\phi_{i+1}, a'; \theta) & \text{For non-terminal } \phi_{i+1} \end{cases}$$
- Utilize a gradient descent step on $(y_i - Q(\phi_i, a_i; \theta))^2$

   end for

   end for

FIG. 2 The pseudocode for the Q-learning algorithm.

The optimum demand of prosumer p at time slot t is yield from the following [28]:

$$d_{p,t} = \arg \max \omega_{p,t}\left(d_{p,t}\right) = \begin{cases} d_{p,t}^{\min} & \text{if } \dfrac{\beta_{p,t} - \pi_t}{\gamma_{p,t}} < d_{p,t}^{\min} \\[2ex] \left(\beta_{p,t} - \pi_t\right)/\gamma_{p,t} & \text{if } d_{p,t}^{\min} < \dfrac{\beta_{p,t} - \pi_t}{\gamma_{p,t}} < d_{p,t}^{\max} \\[2ex] d_{p,t}^{\max} & \text{if } d_{p,t}^{\max} < \dfrac{\beta_{p,t} - \pi_t}{\gamma_{p,t}} \end{cases} \quad (1)$$

in which π_t is the market price and the local market price at time slot t is represented by $d_{p,t}$. $\beta_{p,t}$ and $\gamma_{p,t}$ are the parameters of the proposed Q-learning algorithm that are time dependent.

Therefore the optimal price, $\pi_{p,t}$, and demand, $d_{p,t}$, of each prosumer at t would be determined through the Q-learning algorithm. As aforesaid the offered generation power of each prosumer equals its maximum capacity predicted at the previous hour, denoted by $\rho_{p,t}$. Prosumers submit their offered demand, price, and generation capacity to the local market via their blockchain account.

In the next step, demand and its corresponding prices will be aggregated in a descending order, whereas renewable capacities and the corresponding prices are aggregated in an ascending fashion, i.e., the generation capacity with lower prices and the demand offered the higher prices are the first priority. The intersection of the line chart of the aggregated demand and the aggregated supply determines the unique market price for the time slot t, π_t. Fig. 3 illustrates the clearing price for the hypothesis situation.

In the final step of stage 1, the DR price, π_t^{DR}, is determined by Eq. (2):

$$\pi_t^{DR} = \begin{cases} 0 & \text{if } \displaystyle\sum_{p=1}^{P} d_{p,t} \leq \sum_{p=1}^{P} \rho_{p,t} \\[3ex] \alpha_t \pi_t & \text{if } \displaystyle\sum_{p=1}^{P} d_{p,t} > \sum_{p=1}^{P} \rho_{p,t} \end{cases} \quad (2)$$

where in Eq. (2), $\alpha_t \geq 1$. α_t is a time-dependent parameter specified by the system operator. It is assumed that the local market has P prosumers. Eq. (2) states that the market would incentivize prosumers to reshape their load if the total demand of the market cannot be met by the local supply. Note that Eq. (2) should be subjected to the following constraint:

$$\pi_t^{DR} \leq \pi_{market,t} \quad (3)$$

In Eq. (3), $\pi_{market,t}$ is the price of the upstream market at t. The constraint states that the DR price would not be profitable if it exceeds the price of the upstream market.

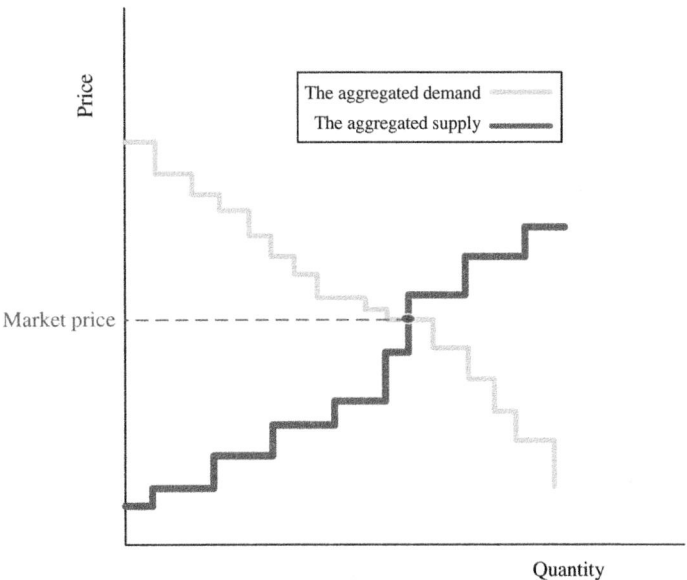

FIG. 3   The first settlement of the local market.

Consequently the prosumers can receive the market price and DR price via their blockchain account. Then the next stage begins.

4. Stage 2: The prosumer's rescheduling and the final settlement

Having received the actual price and the DR price signals, the prosumers of the local market are given the chance to reschedule their demand, using Eq. (1). The amount of demand offered by prosumer p to be shifted from t or curtailed during t is called the DR part of the demand, denoted by $d_{p,t}^{DR}$, given from the following:

$$d_{p,t}^{DR} = d_{p,t} - d_{p,t}^{new} \tag{4}$$

where $d_{p,t}^{new}$ denotes the demand rescheduled by the prosumer at t in the second stage.

Finally the market is settled aiming at maximizing the social welfare of the local market:

$$\max \left(\sum_{p=1}^{P} \pi_t \rho_{p,t}' + \sum_{p=1}^{P} \pi_t^{DR} d_{p,t}^{DR'} - \sum_{p=1}^{P} \pi_t d_{p,t}^{new'} \right) \tag{5}$$

Subject to

$$0 \leq \rho_{p,t}' \leq \rho_{p,t} \tag{6}$$

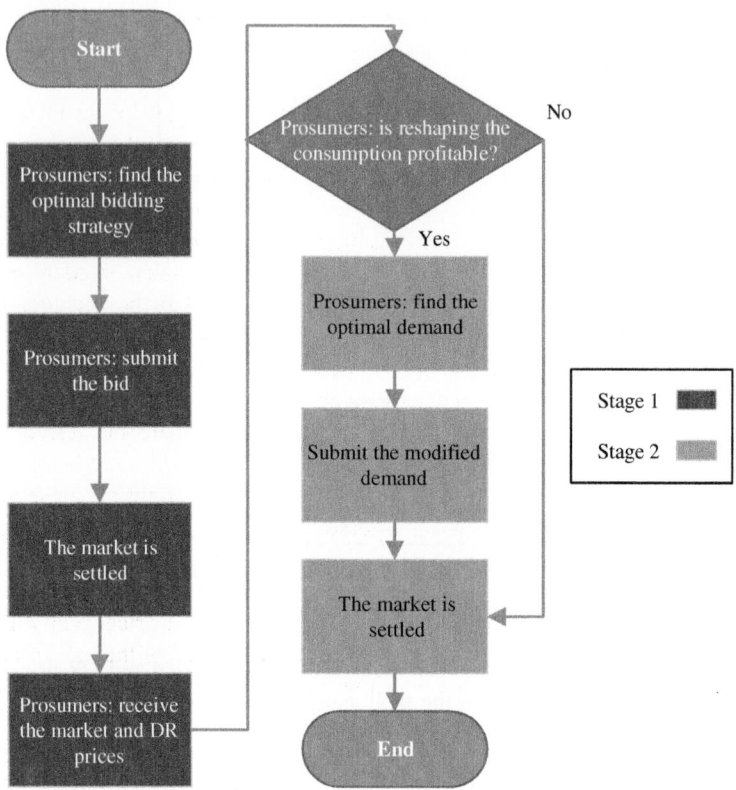

FIG. 4   The flowchart of the proposed model.

$$0 \le d_{p,t}^{DR'} \le d_{p,t}^{DR} \tag{7}$$

$$0 \le d_{p,t}^{new'} \le d_{p,t}^{new} \tag{8}$$

where $\rho_{p,t}'$, $d_{p,t}^{DR'}$, and $d_{p,t}^{new'}$ are the optimum amount of renewable generation, DR demand, and rescheduled demand of customer p at t, respectively, which should be determined from Eqs. (5)–(8). Constraints (6), (7), and (8) explain that the optimal amount of renewables, DR demand, and rescheduled demand should not exceed the values offered by the prosumers at t. Note that the optimal market price and the DR price were extracted from the settlement of the previous stage.

Fig. 4 depicts the flowchart of the proposed model. The first and second stages are specified in the figure.

5.   The blockchain-based platform

In this chapter, each prosumer is regarded as an independent agent that would be able to have access to the blockchain platform. The system operator is also responsible for registering the qualified prosumers as agents to take part as participants. Furthermore the operator determines the optimum amount of α_t for each time slot of the day considering the market prices.

The implementation of the local market is built on a private blockchain so that the prosumers and those customers who have flexible demand can access the local market. Each participant has a unique address in the chain that is connected to the checking account on the smart contract [29]. Hence the participant can deposit and withdraw money through its blockchain account. The market rules, instructions, and mechanisms related to the settlement are implemented by one of the important features of blockchain, the smart contract. In fact the smart contract provides the link between prosumers intending to buy power and those who want to sell it. In the final settlement the account balances of participants will be settled after the offered prices and quantities of buyers and sellers are settled. Finally the transactions are executed, new blocks are built, and the funds are transferred from the accounts of the sellers to those of buyers.

Moreover the penalty would be deducted from the account of the seller if there exist differences between the offered capacity and the actual capacity shared with the market. This can be due to the error of forecast or variability resulted from the intermittent characteristic of renewables and demand. The amount of the penalty mainly depends on the local market prices and will be determined by the system operator. Note that if the demand of the local market cannot be met through its own supply, it will be supplied by the upstream market. Therefore the prices of the upstream market would also affect the penalty.

The local markets definitely benefit from disintermediation, which blockchain can bring. The elimination of the central profit-based entity can considerably decrease the prices of the local market. In addition, blockchain technology builds a single venue in which a number of participants can compete with each other to sell and buy energy, ensuring the best possible prices in the local market. So the utilization of blockchain technology is more cost-efficient [30]. Each transaction record is transparent and open through the use of a blockchain platform. A prosumer can have unrestricted access to the previous transactions in which it was involved. Moreover the privacy and security of input data are ensured using the hash functions. To this end, it promotes trust among market participants. They can share data and energy with each other within the local market while staying anonymous.

6.   Simulation: Numerical results

6.1   Decentralized versus centralized aggregator-based scheme

In this section, we aim to analyze the effects of decentralization on the prices of the market. To this end, first, the results extracted from the proposed model will be compared with the results of the model that has an aggregator. In this model the aggregator determines the billing costs of its downstream customers and prosumers, whereas in the proposed model the prices paid by prosumers are specified through the local market settlement. In the aggregator-based scheme, the aggregator collects all of the information about the prosumers, aiming to maximize the prosumers' profits from buying and selling power. It will be based on the incentive-based DR program in which the aggregator and prosumers reach a contract so that the aggregator can manage the prosumers' generation and consumption.

To capture uncertainties of the upstream market, different scenarios were generated through the use of the roulette wheel mechanism according to the historical data of the Australian electricity market. Then the prices of the proposed local market are compared with the billing costs of the same prosumers who are aggregated by the aggregator. The simulation has been done for two subsequent days, the first and the second days of a week. Figs. 5 and 6 illustrate the comparison between the prices of the proposed model and the aggregator-based one for the first and the second days of a week, respectively. Note that the aggregator is considered as a nonprofit entity.

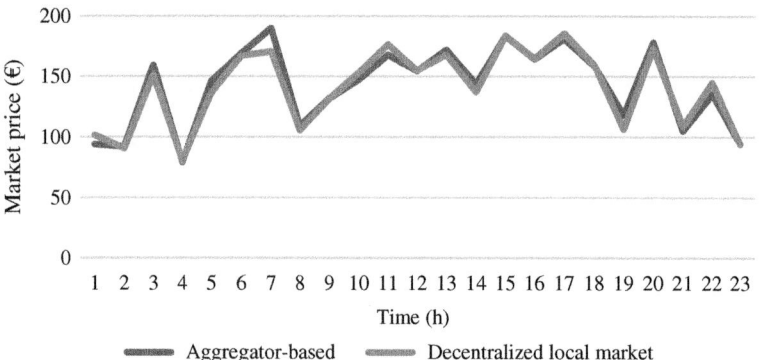

FIG. 5 The results of decentralized proposed and the aggregated-based one for the first day of a week.

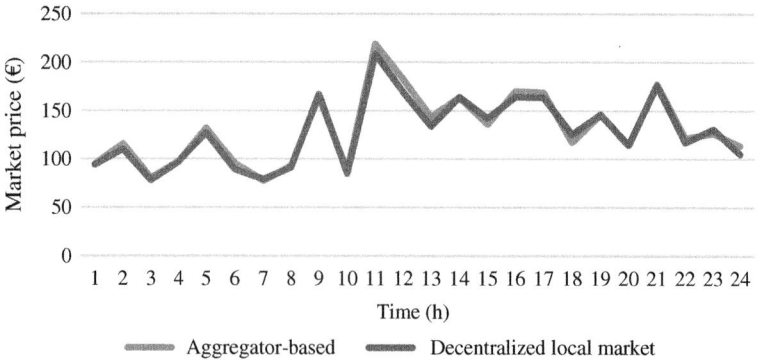

FIG. 6 The results of decentralized proposed and the aggregated-based one for the second day of a week.

According to the results, although the aggregator-based model requires some private information on all of the prosumers, the results are almost the same. In other words the simulation states that prosumers are not required to share their information with the aggregator to minimize their billing costs. They can preserve their privacy such as their comfort cost, have the freedom to bid their demand and generation capacity, and preserve their privacy using the proposed decentralized model.

6.2 The effects of the contribution of the prosumers on the market load profile

In the second part of the simulation, we compare the total demand of the prosumers in the first and the second stages. In the first stage, in which participants did not receive the price signals, they would submit offered demand aiming to maximize their comfort. Figs. 7 and 8 show the optimal market prices and the

FIG. 7 The local market prices for 24h of a day.

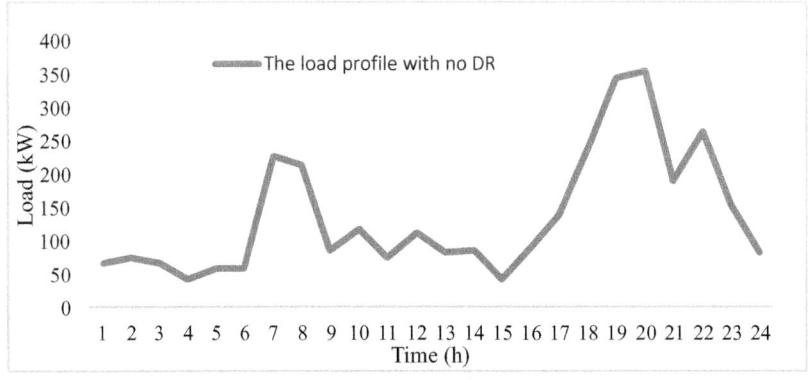

FIG. 8 The aggregated demand for prosumers in the first stage.

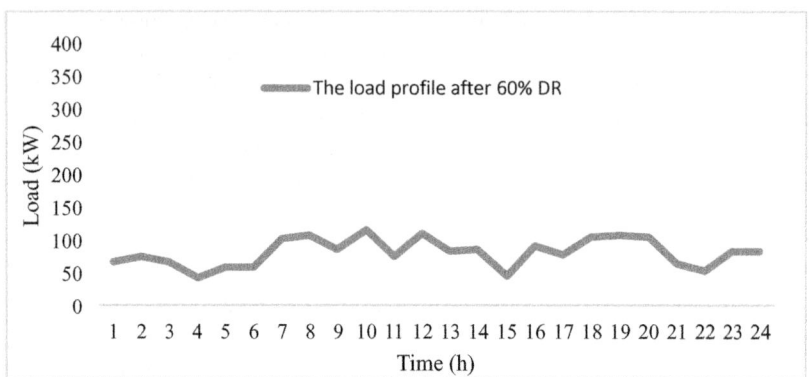

FIG. 9 The aggregated demand of prosumers in the second stage after 60% DR participation.

total demand for 24 h of a day, respectively. The figures state that the trends of the load profiles and the market prices are the same. In other words the market prices will increase when the demand of the participants is growing.

However, after receiving the signals in the second stage, it is assumed that 60% of prosumers decide to reshape their demand. The new load profile of the market with 60% of prosumers' contribution is shown in Fig. 9. As the figure denotes, the prosumers' demand follows an approximately monotonous trend in comparison with the previous stage. They have modified their demand, especially in peak time slots so as to minimize their costs during the day. Even though in the proposed study the potential LV network constraints were not taken into account, as it can be seen from the simulation part of reference [31], how the load profile of prosumers is flattened due to DR participation is also beneficial from potential distribution (LV) network constraints point of view.

7. Conclusion

This chapter introduces a two-stage model to obtain a decentralized price-based demand response program. In the proposed model the privacy of prosumers is preserved since there exists no entity acting as an intermediary. Hence the prosumers are not required to share their information with another entity such as an aggregator. The first stage of the model starts with the scheduling of the prosumers using Q-learning method. The prosumers submit the optimal bids to the local market, the market is settled, and the optimal market and demand response prices are sent to the prosumers. Then the prosumers can opt to change their consumption in response to the prices they received. The final settlement of the market will be reached after receiving the modified demand. The smart contract, the important feature of blockchain technology, is utilized to execute transactions and implement the proposed market mechanism. In addition, the proposed decentralized model is compared with the centralized aggregator-based one. The results state that prosumers can have approximately the same profits in the decentralized demand response program while preserving their own privacy. In addition, the load profile of the market can follow an approximately consistent trend if 60% of participants choose to reschedule their demand to react to the price signals. This trend can be also beneficial from potential distribution network point of view.

References

[1] G. Strbac, Demand side management: benefits and challenges, Energy Policy 36 (12) (2008) 4419–4426.

[2] J.A. Ruiz-Arias, J. Terrados, P. Pérez-Higueras, D. Pozo-Vázquez, G. Almonacid, Assessment of the renewable energies potential for intensive electricity production in the province of Jaén, southern Spain, Renew. Sust. Energ. Rev. 16 (5) (2012) 2994–3001.

[3] M. Hinnells, Technologies to achieve demand reduction and microgeneration in buildings, Energy Policy 36 (12) (2008) 4427–4433.

[4] C. Delmastro, E. Lavagno, G. Mutani, Chinese residential energy demand: scenarios to 2030 and policies implication, Energy Build. 89 (2015) 49–60.

[5] H.T. Haider, O.H. See, W. Elmenreich, A review of residential demand response of smart grid, Renew. Sust. Energ. Rev. 59 (2016) 166–178.

[6] P. Siano, Demand response and smart grids—a survey, Renew. Sust. Energ. Rev. 30 (2014) 461–478.

[7] R. Earle, A. Faruqui, Toward a new paradigm for valuing demand response, Electr. J. 19 (4) (2006) 21–31.

[8] L. Gkatzikis, I. Koutsopoulos, T. Salonidis, The role of aggregators in smart grid demand response markets, IEEE J. Sel. Areas Commun. 31 (7) (2013) 1247–1257.

[9] A. Soares, Á. Gomes, C.H. Antunes, Categorization of residential electricity consumption as a basis for the assessment of the impacts of demand response actions, Renew. Sust. Energ. Rev. 30 (2014) 490–503.

[10] F. Shariatzadeh, P. Mandal, A.K. Srivastava, Demand response for sustainable energy systems: a review, application and implementation strategy, Renew. Sust. Energ. Rev. 45 (2015) 343–350.

[11] M. Nikzad, B. Mozafari, Reliability assessment of incentive-and priced-based demand response programs in restructured power systems, Int. J. Electr. Power Energy Syst. 56 (2014) 83–96.

[12] Y. Wang, Q. Chen, C. Kang, M. Zhang, K. Wang, Y. Zhao, Load profiling and its application to demand response: a review, Tsinghua Sci. Technol. 20 (2) (2015) 117–129.

[13] M.A. López, S. De La Torre, S. Martín, J.A. Aguado, Demand-side management in smart grid operation considering electric vehicles load shifting and vehicle-to-grid support, Int. J. Electr. Power Energy Syst. 64 (2015) 689–698.

[14] A.J. Roscoe, G. Ault, Supporting high penetrations of renewable generation via implementation of real-time electricity pricing and demand response, IET Renew. Power Gener. 4 (4) (2010) 369–382.

[15] Z. Zhou, F. Zhao, J. Wang, Agent-based electricity market simulation with demand response from commercial buildings, IEEE Trans. Smart Grid 2 (4) (2011) 580–588.

[16] P. Faria, Demand Response in Future Power Systems Management—A Conceptual Framework and Simulation Tool, Instituto Politécnico do Porto, Instituto Superior de Engenharia do Porto, 2011.

[17] A. Moshari, A. Ebrahimi, Reliability-based nodal evaluation and prioritization of demand response programs, Int. Trans. Electr. Energy Syst. 25 (12) (2015) 3384–3407.

[18] H. Khajeh, A.A. Foroud, H. Firoozi, Robust bidding strategies and scheduling of a price-maker microgrid aggregator participating in a pool-based electricity market, IET Gener. Transm. Distrib. 13 (4) (2018) 468–477.

[19] C. Vivekananthan, Y. Mishra, G. Ledwich, F. Li, Demand response for residential appliances via customer reward scheme, IEEE Trans. Smart Grid 5 (2) (2014) 809–820.

[20] M.A.F. Ghazvini, J. Soares, N. Horta, R. Neves, R. Castro, Z. Vale, A multi-objective model for scheduling of short-term incentive-based demand response programs offered by electricity retailers, Appl. Energy 151 (2015) 102–118.

[21] J. Vuelvas, F. Ruiz, G. Gruosso, Limiting gaming opportunities on incentive-based demand response programs, Appl. Energy 225 (2018) 668–681.

[22] M. Yu, S.H. Hong, Incentive-based demand response considering hierarchical electricity market: a Stackelberg game approach, Appl. Energy 203 (2017) 267–279.

[23] M. Yu, S.H. Hong, Y. Ding, X. Ye, An incentive-based demand response (DR) model considering composited DR resources, IEEE Trans. Ind. Electron. 66 (2) (2018) 1488–1498.

[24] S. Bahrami, M.H. Amini, M. Shafie-khah, J.P.S. Catalao, A decentralized electricity market scheme enabling demand response deployment, IEEE Trans. Power Syst. 33 (4) (2017) 4218–4227.

[25] S. Bahrami, M.H. Amini, M. Shafie-Khah, J.P.S. Catalao, A decentralized renewable generation management and demand response in power distribution networks, IEEE Trans. Sustain. Energy 9 (4) (2018) 1783–1797.

[26] H. Khajeh, A.A. Foroud, H. Firoozi, Optimal participation of a wind power producer in a transmission-constrained electricity market, in: 2017 25th Iranian Conference on Electrical Engineering, ICEE 2017, 2017.

[27] M. Shafie-khah, G.J. Osório, J.P.S. Catalão, A decentralized privacy-based market scheme for responsive demands, in: 2017 IREP, August 27–September 1, Espinho, Portugal, 2017.

[28] B. Chai, J. Chen, Z. Yang, Y. Zhang, Demand response management with multiple utility companies: a two-level game approach, IEEE Trans. Smart Grid 5 (2) (2014) 722–731.

[29] E. Mengelkamp, J. Gärttner, K. Rock, S. Kessler, L. Orsini, C. Weinhardt, Designing microgrid energy markets: a case study: the Brooklyn Microgrid, Appl. Energy 210 (2018) 870–880.

[30] P. Ellis, J. Hubbard, Flexibility trading platform—using blockchain to create the most efficient demand-side response trading market, in: A. Marke (Ed.), Transforming Climate Finance and Green Investment With Blockchains, Elsevier, 2018, pp. 99–109.

[31] J. Guerrero, A.C. Chapman, G. Verbič, Decentralized p2p energy trading under network constraints in a low-voltage network, IEEE Trans. Smart Grid 10 (5) (2019) 5163–5173.

Chapter 8

Blockchain in decentralized demand-side control of microgrids

Amin Hajizadeh[a] and Seyed Mahdi Hakimi[b]

[a]*Department of Energy Technology, Aalborg University, Esbjerg, Denmark,* [b]*Department of Electrical Engineering, Damavand Branch, Islamic Azad University, Damavand, Iran*

1. Introduction

Real-time control and supervision participate in an important part in the management of smart energy networks and process at medium- and low-voltage levels. Newly, as a result of the quick development in the arrangement of Distributed Energy Prosumers (DEPs), the smart grid management problems using centralized methods can no longer be proficiently; therefore the requirement for decentralized methods and structures is extensively identified [1–4]. The expansion of Internet of Things (IoT) and smart metering devices together with the vision of renewable energy incorporation has improved the level of implementation of decentralized energy networks where, owing to the absence of grid energy storage capacity, electrical energy must be used as it is generated [5]. However, the integration of renewable energy has enhanced a level of uncertainty as a result of the irregular and unpredictable characteristics of its resources [6]. Output changes in energy production may threaten the safety of energy supply, resulting in overloading of energy parts and power outages or disruptions in service. In some situation, because of an unexpected peak of renewable energy generation in the smart grid, the energy demand and energy production are not balanced. The energy demand is inadequate to meet the whole produced energy. The problem is intensified by the lack of capabilities of DSOs to frequently decrease the output of energy production sources not to compromise the entire grid operation. A proper method for these problems is the demand-side management (DSM) targeting at matching the energy demand with the production by inspiring DEPs to shed or shift their energy demand to deal with peak load periods [7,8]. In this system the DSOs have characterized demand response (DR) programs giving the prospect to DEPs to assume a

Blockchain-based Smart Grids. https://doi.org/10.1016/B978-0-12-817862-1.00008-7

critical job in the task of the power framework by forming their vitality request to meet different lattice level objectives and get in return money-related advantages [9]. Regularly the DSO starts a DR occasion toward the start of a charging period by sending a guideline sign to each DEP determining a solicitation to adjust the utilization for a restricted period and the money-related motivations [10, 11]. The DEPs send offers with the measure of vitality, and they are happy to lessen or to expand their interest, while the DSO acknowledges the offers and checks if the harmony between the absolute vitality request and age at network level is met. Subsequently the DEPs will intentionally plan their activity for gathering the concurred profiles by time moving a few errands that require some measure of electric vitality or by changing piece of their utilization to substitute sources. In this way, DR projects offer a few advantages to the vitality frameworks, including expanded effectiveness of benefit use and more prominent entrance of renewables without diminishing dependability, facilitating limit issues on circulation systems to encourage further take-up of dispersed age on clogged neighborhood systems, decreasing the required generator edge and expenses of approaching conventional hold, and including the related ecological advantages through diminished outflows [12]. Over the most recent couple of years, the scholarly, inquire about, and mechanical spaces have picked up a great deal of enthusiasm for the dispersed record and blockchain innovation and its potential in decentralizing the administration of complex frameworks. The conveyed record [13] is made from a set out of squares, chained back utilizing a connected rundown of hash pointers, each square putting away a lot of legitimate exchanges of computerized resources. The connected rundown is an affix just information structure; in this way, any progressions that would show up in past enlisted squares would prompt irregularities (i.e., the hash pointer of that square would change). On the off chance that one needs to change the substance of a past square, all the accompanying ones should be reiterated and connected to acquire a steady refreshed information structure. The favorable position brought by this structure is the sealed log of all value-based data contained in the squares. Every one of the exchanges and squares is disseminated (i.e., recreated) among the hubs of a distributed system. Enrolling another exchange will send it to all its companion hubs, and every one of them will approve and proliferate it further. On the off chance that any logical inconsistencies or invalid states happen, the exchange won't be sent. To maintain a strategic distance from circles in the system, a hub can choose not to advance exchange on the off chance that it was at that point recently enrolled. Since there is no focal expert to make new squares and every hub keeps a neighborhood duplicate of the record, accord calculations are utilized to guarantee that every one of the hubs concurs upon a worldwide truth about the substantial record state. The accord calculations, for the most part, depend on proof protocols [14] that characterize computational serious issues that are hard to unravel and moderately simple to approve. Another square containing the most current exchanges distributed in the system is mined and approved by a hub that finds an answer for that issue (see green square in Fig. 1).

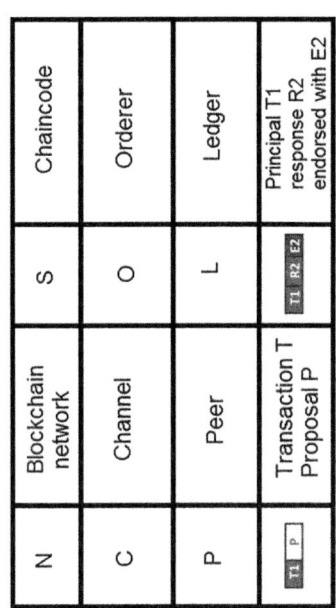

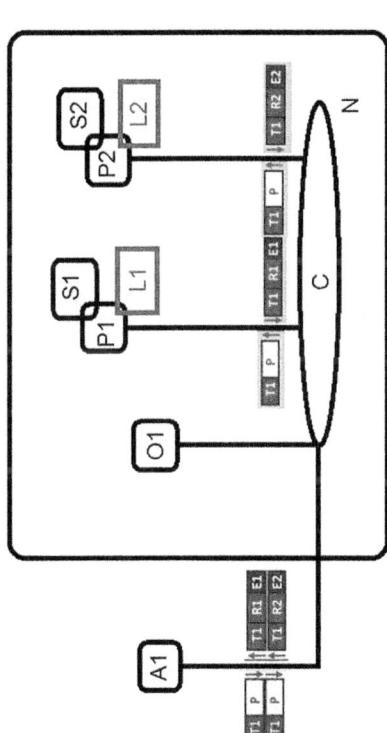

FIG. 1 Ledger distribution for peer-to-peer network [14].

New forms of blockchain innovation usage offer help for the execution of shrewd contracts [13]. The brilliant contracts are bits of code that execute diverse business decisions that should be checked and settled upon by all companion hubs from the system. These agreements are enrolled in record's squares and activated by exchange calls that require every hub to refresh its state dependent on the outcomes got in the wake of running the keen contract. Since they are additionally repeated in every one of the hubs of the system, they offer incredible potential for control decentralization. They go about as specialists that can have a state and usefulness and can be activated anytime in the wake of being effectively conveyed, supplanting outsider center elements from the value-based world (judges, litigators, escrows, and so on). In our vision the blockchain innovation can possibly give a troublesome inventive way to deal with DR projects and vitality exchanges, making ready for verified cryptography-based decentralized administration of practicality vitality lattices. In this specific situation the fundamental commitments of this book section are the following:

- a blockchain-based model for distributed management, control, and validation of DR events in low-/medium-voltage smart grids;
- Blockchain-based distributed ledger for storing the data acquired from metering devices as energy transactions in a secure and tamper-proof manner;
- Implementation of self-enforcing smart contracts to track and check the compliance of each DEP enrolled in DR programs to the desired demand energy profiles, to calculate associated rewards and penalties, and to detect grid energy unbalances requiring the definition of new DR events;
- Finally a consensus-based DR validation approach to activate the appropriate financial settlement to the flexibility providers and to increase the reliability of the smart grid operation.

2. Related work

Various studies have been done on the blockchain algorithm in which this paper focused on distributing the energy of the hybrid energy via the blockchain; moreover the case study shows that the electricity consumption is tracking the pattern of the renewable generation with P2P method of exchange [1]. The other important factor is the bandwidth demand, as the requirement for the blockchain is 10 times higher than the demand in real-time AMI in both standard and high scenario conditions; thus a higher bandwidth for communication is a must for the blockchain transactions [2]. The energy management of residential, commercial, and industrial users plays a vital role; therefore three parameters were considered for it, which are technical issues such as high frequency of P2P power trading, decentralizing application for P2P power trading on blockchain, and developing and motivating the application for attracting more participants [3]. A new model for decentralized transactive energy management leads to the optimization of energy and financial flows in the transition toward active distribution networks [4].

In this book chapter, old-fashioned charging was replaced by mobile charging vehicle (MCV); in addition to that an optimal scheduling framework was implemented with a consortium blockchain to achieve minimum cost and maximum satisfaction. The deed was done for one of the ring roads of Beijing, and the proportions of charging show better performance on cost efficiency and user satisfaction [5]. The blockchain can be used for electrical vehicles (EV) and the required renewable energy for them since in urban districts power consumption and a conventional scheme could make the energy optimization much better [6]. Another recent paper is that the trust consensus protocol is proposed to reach an agreement on electric vehicles with blockchain, in which credibility based on trust and validation calculations is being proposed [7]. This paper proposes a project for the adaptive blockchain-based electric vehicle (AdBEV). The AdBEV design uses the Iceberg order execution algorithm to match the smart grid demand and charge demand [8].

This paper emphasizes on DSM systems, which are a key tool in improving collective self-consumption. Therefore, by using the optimization methods for DSM and with the rising of blockchain technology, new opportunities, such as consumption and production measurements validation, and local energy market implementation were discussed [9].

The paper first uses the benefits of blockchain technology to provide a P2P secure trading framework among ESS about ADR programs without relying on a trusted third party [10].

Electric vehicles act as power storage devices for charging operations at load times, and their energy is fed back to the grid to reduce the maximum load. Vehicles can also sell their energy to adjacent charging vehicles in a P2P-enabled way with local collectors; for instance, in autonomous driving, the need to increase the accuracy of data analysis and make a quick decision is essential [11].

This paper is a system that focuses on demand management in the network and blockchain in industrial environments using machine-to-machine protocols (M2M) [12].

This paper proposes a game-based theory-based technique for managing the demand side while combining storage systems with regard to supply constraints. The proposed game-based approach not only reduces the peak and mean (PAR) but also balances the division of profits and demand, as well as blockchain technologies that are introduced to secure the implementation of an extended approach [13].

Here is a conducted review to develop an innovative framework for empowering consumers and customers using blockchain-based P2P microprocessors [14].

Blockchain can be used to achieve authentication, authorization, accountability, security, integrity, compatibility, and nonrejection for real-time applications, which may be provided by an efficient centralized system in a smart social environment; the authors devised a blockchain-based framework to support the right energy business in a network-based vehicle system recently [15].

This paper presents a blockchain-based architecture for managing DR distribution, control, and validation in low-/medium-voltage smart grids, thus ensuring high reliability and decentralized operation through reliable and flexible trading transactions and reliable and real-time DR validation. To validate the proposal a simulation prototype was implemented using the Ethereum platform. The results showed that blockchain-based distributed demand-side management can be used for energy and production demand in an intelligent grid, and the demand response signal is followed with high precision, while the amount of energy flexibility required for convergence is reduced. However, the proposal does not consider multidimensional markets [16].

3. Blockchains and renewable energies

3.1 Blockchain basics

Blockchains run on digital networks. Data communication in such networks is equal to copying information from one position to another; for example, in the cryptocurrency domain, this is equivalent to copying digital coins from one user's electronic wallet to another. The main challenge resides in the fact that the system wants to make sure that coins are only spent once, avoiding twice spending. A conventional answer is to use a central point of authority, such as a central bank, who acts as the confidence agent between transacting parties. On some occasions, central managing may not be possible or pleasing, as it initiates intermediary costs and needs network users to trust a third party to work the system [17]. Centralized systems also have important inconveniences due to a single point of failure, which renders them additionally vulnerable to both technical failures and malicious attacks [18]. The main principle of blockchain technologies is to eliminate the need for such intermediaries and substitute them with a distributed network of digital users who work in partnership to confirm transactions and ensure the integrity of the ledger. If central management is removed, the challenge resides in finding a professional method to consolidate and synchronize various copies of the ledger. The accurate procedure of validation and ledger consolidation varies for different types of blockchains. These support mechanisms are known as distributed consensus algorithms [19]. Blockchains can be public or private; the only difference is related to who can participate in the network [20]. There are different protocols of agreement in blockchain technology. These are policies that every network uses to confirm information. The method to achieve consensus is essentially related to transaction velocity, safety, transparency, and scalability. With PoW, the majority well-known consensus algorithm, used by Bitcoin, miners compete to add a novel block to the existing blockchain by solving a problem. Miners have no technique to forecast or influence the product, so the only possible achievement is that of trial and error. This brute-forcing process requires computational attempt and therefore electricity. When the problem is solved, the block is returned to the Bitcoin network and is accepted by other nodes if all transactions are suitable and unspent, and the winning miner takes a financial prize. By starting work on the consecutive block, other miners accept the recently generated block. Significantly, all succeeding blocks have puzzles solved from all preceding blocks. As the generation of novel puzzles is accidental and performed in parallel by a lot of miners, several chains may appear. In this time the network stores all resulting chains. Network members eventually abandon all other chains but the longest, which is assumed to have been produced by a network majority of computational power and therefore represents the most valid state of the ledger. As a result, malicious attackers are constantly outpaced by the honest part of the network, unless they can control more than 51% of the computational power in the network. In the case of a 51% attack, malicious nodes could potentially rewrite the entire history of transactions. One of the disadvantages of PoW is the computational power needed to carry out the tasks to confirm the transactions, something that requires enormous amounts of electricity. Sources report that Bitcoin could use large amount of electricity in Denmark by 2020. On an extra positive note, a new study made by CoinShares concluded that Bitcoin procured 77% of its energy use from renewable energy. PoS is yet one more method to confirm a transaction. It aims to reach consensus by replacing the brute force of computational power and energy use with an accidental collection procedure depending on the wealth of each of the participants or node owners [21]. This makes the blockchain achieve consensus a lot quicker and smaller amount of energy intense. The rewards are different than novel coins. Instead, they only take transaction fees. Ethereum, one of the majority well-known blockchain platforms, is contemplating the shift from PoW to PoS [22]. This means changing the software protocol that supports the blockchain by the participants involved. This update is called a hard fork and is meant for decreasing the number of rewards given, reducing inflation pressure in the cryptocurrency. Given the energy demand of a PoW approach, several developers are showing preference for other consensus algorithms such as PoAu. The block generation in PoAu requires granting special permission to one or more members to make changes in a blockchain. Network members put their trust into authorized nodes, and a block is accepted if many authorized nodes sign the block. Any new validator can be added to the system via voting. Although the method represents a more centralized approach, most appropriate for governing or regulatory bodies, it is currently also proving popular with utilities in the energy sector. An example is the Energy Web blockchain that will run on a proof-of-authority algorithm named Aura [23]. With the growth of digitalization in approximately all parts in personal and business life, a lot of novel possibilities release for optimizing processes and create them smoother. In the whole world, there is a development toward replacing outdated paper registries and databases with modern, digitized systems. This makes information easier to discover and distribute. Blockchain technology is a novelty that promises efficiency gains in business-to-business support and in daily transactions. It can assist the interaction between parties without them having to believe each other and, thanks to the immutability of its transaction history, suggest a secure platform for information that is to be stored securely. By its distributed construction, blockchain technology can also make simpler the collaboration between some parties and build it safer and more transparent. Based on these characteristics, blockchains are extremely fine matched for developing supply chains and equipping them for an increasingly worldwide and flexible world. Blockchain's ability to allow peer-to-peer energy transactions could significantly disrupt the energy sector, particularly by encouraging decentralization. The growing use of small renewable energy installations, such as rooftop solar panels, can create stress on electricity grids that were designed with large, centralized power plants in mind. By allowing peer-to-peer energy trading and incentivizing local consumption at the time of production, blockchain could stabilize the grid, aiding this decentralization. However, with users paying each other directly, many of the traditional market roles could be called into question, including distribution system operators, retailers, suppliers, metering point operators, and balancing groups. Pilot projects for community energy and peer to peer have already been successfully run by the Brooklyn Microgrid in New York, PowerLedger in Australia, Conjoule in Germany, and many more. However, in Europe, these experiments are limited to pilots under regulatory exemptions or private microgrids—peer to peer remains far from being rolled out universally. Blockchain could also be used for electricity tracking with at least two purposes: rewards for generating renewable energy (e.g., SolarCoin) and renewable energy certificates or carbon credits. For those who want to invest in renewables but lack the funds, blockchain technology could enable collective investments, ensuring fair and transparent sharing of revenues. Electricity systems are being challenged by the introduction of high volumes of renewable energy generation from decentralized sources that demand for new tools to maintain safe operation and stability. Also the electricity sector is on the edge of digitalization with the deployment of sensors and smart devices at the premises of every consumer in numerous countries. There is a growing interest in blockchain technologies in the electricity sector because blockchain enables distributed transactions with transparency and immutability. Therefore it is an ideal technology to face the challenges of decentralized generation systems. Blockchain technology is the union of different technologies such as cryptography, P2P networks, and data ledgers. The most famous use of blockchain is Bitcoin. Bitcoin was born in 2008 with other cryptocurrencies appearing thereafter with different applications. According to a Gartner report, the peak of inflated expectations already passed for blockchain technologies. The report states that all emerging technologies transit between different stages in the hype cycle, from innovation trigger and peak of inflated expectations where hype is at its maximum to the valley of disappointment and at last the plateau of productivity. Concrete developments will appear only now that hype has passed. One of the clear opportunities of blockchain technology is the energy sector where all major utilities are exploring use cases. The reality is that applications like Bitcoin with a complete decentralization and an expensive infrastructure to maintain are not the best for the electricity ecosystem. Different consensus protocols are proposed. All of them have advantages and challenges ahead. Some of them tackle issues like security and energy consumption in different ways. Numerous developers are working on the use of blockchain technologies for renewable certificates, their automatic issuance, and trading. One of the earliest solar energy certificates on blockchain was solar power certificates developed by Linq platform in 2016. SolarCoin is another example: for every megawatt-hour of solar energy fed into the grid producers is awarded one SolarCoin, which can be either stored in a SolarCoin wallet or converted to bitcoins. SolarCoin is partnered with SMA, a German inverter company, to tap several GW of small- and medium-sized generators around the world. Another relevant example is NRGcoin that was born as an intellectual project and is now continued by Enervalis. The NRGcoin method replaces conventional high-risk renewable support policies with a new blockchain-based smart contract, which better rewards green energy. Although the centralized energy market structure has an inadequate number of decision-makers, decentralized constructions may engage a great number of factors, among which market and business models require to be coordinated, requiring specialized techniques. One example of decentralized constructions is renewable energy communities, especially in Europe and the United States (Fig. 2). In those communities, citizens take collective action in a renewable energy project at different stages with different roles. These decentralized structures are relatively new in Mexico. In such systems, blockchain in combination with emerging fields such as IoT and smart meters, digital wallets, and smartphones can trace energy from generation to consumption, from business to business using the distribution system.

4. Blockchain applications in microgrids

The decentralized construction of blockchain fits into the decentralized approach for control and business processes in a microgrid. In this section, preferred blockchain projects and concepts for microgrids are presented. Most projects are still below expansion or in the testing phase; hence, projects with publicly accessible information were chosen primarily.

(1) PWR Company

PWR Company [24] focuses on P2P renewable energy trading in microgrids. As a replacement for selling the energy instantaneously, PWR equips homes

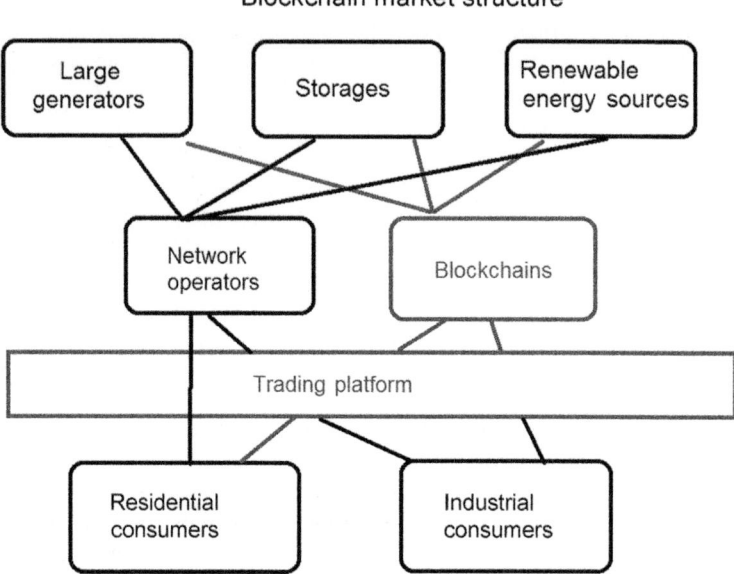

Blockchain market structure

FIG. 2 Proposed structure to sell CECs from distributed energy to an energy supplier [24].

with deep cycle batteries for energy storage to stabilize the grid. The project presently uses the Ethereum platform, which will be replaced in the future by their own version of an energy-based cryptocurrency, the PWRToken. One PWRToken equals to 1 MWh and can be traded on various trade markets.

(2) Power Ledger

Power Ledger provides a market trading and clearing mechanism based on blockchain [25]. Owners of renewable energy resources can sell their surplus of energy at a selected cost inside microgrids or over the distribution grid. Distribution system operators (DSO) receive income for energy traded over the distribution grid.

(3) Key2Energy

In the Key2Energy conception [26], in its place of consuming power from the grid, multiapartment residences present PV energy to its tenants at cheaper costs. In this process, two agents are involved. The first tries to maximize the revenues for the home by selling the produced solar energy on the local market at best probable cost. The second tries to minimize the cost for shared electricity, for example, powering elevators and lighting. The blockchain is used for transactions and smart contracts for the market logic.

(4) LO3 Energy

TransActive Grid and Brooklyn Microgrid LO3 Energy [27] developed the TransActive Grid platform, which is based on Ethereum and smart contracts. The platform aims at different business models for the distribution grid and transactive energy space. It enables peer-to-peer energy transactions, control of DERs for grid balancing, demand response, and emergency management. For this reason, TransActive Grid elements (TAG-e) are developed, which consist of a computer and an electric meter. Their tasks are measuring energy production and consumption, sharing this information with other TAG-e in the network and acting upon this information. The vision is to create a blockchain-based microgrid intelligence system.

(5) Dajie

Dajie [28] presents IoT devices and a blockchain-based platform. To contribute, users have to establish and register one of Dajie's IoT devices. The platform aims at the P2P energy exchange, to apply coins to pay energy and services to energy companies and to redeem carbon credit with coins. Coins can be earned by producing energy. One coin equals 1kWh of produced energy.

(6) Share&Charge

Share&Charge [29] is a network of electric vehicle charging stations. Holders of charging stations can record their stations and set tariffs for charging. Before the registration the station necessity is equipped with the Share&Charge module, which aim is to avoid unauthorized use. Electric vehicle owners can load their Share&Charge wallets with money. The billing of the charging at the station is handled by Share&Charge. Transactions and invoices are stored in the Share&Charge wallet and can be monitored and tracked. The Ethereum platform is used in the transaction layer.

(7) NRGcoin

NRGcoin [30] uses an energy-based cryptocurrency in a structure combining smart contracts. The smart contract structure is based on Ethereum. One NRGcoin is equal to 1 kWh, in spite of the retail worth of electricity. An important difference in this method is that the energy should be produced by renewable energy sources and consumed locally. The smart contract platform is used to process and pay all grid fees and taxes to the DSO. A similar system is responsible to confirm the reported creation of energy by the local clients that not only consume but also produce power. Through the validation phase, it is checked if the energy is consumed locally. If the validation is successful, the produced power is satisfied with NRGcoins, which could be used to pay for future green energy consumption or sold on a currency market. As a result, oversupply in the local area is not rewarded. On the currency market, consumers can buy these

NRGcoins to pay for the green energy; thus they pay 1 NRGcoin per kWh. All kinds of renewable energy are supported, not only solar energy. The operation of NRGcoin measures the electricity flows and communicates with the smart contract and exchange market.

(8) GrunStromJeton

Another still very conceptual structure based on Ethereum is GrunstromJeton [31]. One key component is the use of GrunstromIndex, which is an index that indicates the relative production of energy from alternative, "green" power sources in the next 36 h. When this index is higher, the fraction of power produced from green sources to total energy produced is higher. The system observes the energy consumption of the customers with the use of smart meters and rewards them with GrunstromJetons when they consume power from alternative sources. Therefore the higher the index, the more Jetons the consumer earns, which can be traded and exchanged.

(9) SolarCoin

SolarCoin's aim is to enhance the production of solar energy [32]. Consumers are deterred to invest in solar installations due to long payback times. To reduce it, prosumers are rewarded 1 SolarCoin per produced MWh. With electricity meters, claims for SolarCoins are verified. To register solar installations, different affiliate facilitators exist: SolarChange, ElectriCChain, and SolCrypto.

(10) TheSunExchange

TheSunExchange enables" crowd sale," where users purchase solar cells and lease them to earn a passive income [33]. The project targets developing countries, where government corruption is a major problem. The leasing user pays a rental, from which the solar cell owner receives a Bitcoin income for 20 years. Additionally, the owner earns SolarCoins for the lifespan of the project.

(11) Bankymoon

Bankymoon [34] offers prepaid meters, which are blockchain aware. The idea is to enable funding of electricity, water, and gas to everybody in the world. The meters can be "loaded" by sending payments to the meter in different cryptocurrencies. The pilot project Usizo focuses thereby on needs of African schools, where users around the world can directly spend cryptocurrencies such as Bitcoins to the school's meter to fund, for example, electricity for a month.

5. Blockchain-based management of smart energy grids

A blockchain-based design for disseminated the board, control, and approval of DR programs in low-/medium-voltage savvy networks has been proposed (see Fig. 3) with a perspective on guaranteeing high unwavering quality and decentralized activity by executing identifiable and carefully designed vitality

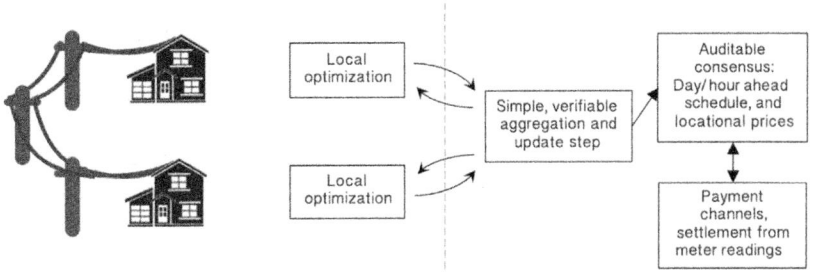

FIG. 3   Structure for decentralized management of energy grids based on the blockchain [14].

adaptability exchanges and close ongoing DR approval. The model the grid is illustrated as a graph of peer nodes (i.e., DEPs; DSOs; and other interested stakeholders such as transmission system operators (TSOs), retailers/suppliers, and aggregators) able to coordinate through a blockchain-based infrastructure to support fully decentralized energy demand and generation matching to ensure stable grid operation.

A blockchain disseminated record is developed and oversaw at the brilliant network level. Each DEP highlights IoT-based vitality metering gadgets and registers the observed information with respect to the vitality generation or vitality utilization esteems in squares as a feature of the record. Along these lines a DEP is displayed as a hub of the shared circulated vitality arrange and can keep up a duplicate of the record that is consequently refreshed when new vitality information is enrolled. One of the real hindrances in DEP commitment in DR projects is information protection and security, which for this situation are inventively tended to utilizing the blockchain appropriated record highlights. The account of vitality exchanges in a carefully designed way is as yet an open research issue in the unified methodologies, despite the fact that a ton of work was put into guaranteeing protection safeguarding keen metering. Endeavors are focusing on giving a trustful bidirectional association between the DEPs and DSO utilizing conventions; for example, OpenADR, however, framework centralization is a key plan flaw that makes information security a fairly touchy issue [35].

In the decentralized approach, energy data are registered and stored locally in the blockchain using DEP digital identities and then replicated and shared with all the network peers for validation.

This is reliable with concentrates on protection and security sought after reaction vitality frameworks that have demonstrated the DEP inclination to keep information in-home and not in unified information storehouses [5]. Exchange permanence is another advantage brought by the blockchain innovation, guaranteeing that any vitality information enlisted in the blockchain stays unaltered after its approval. Since the conveyed record squares are anchored back utilizing a connected rundown of hash pointers, changing the estimation of a vitality exchange in a square (by an aggressor) ends up more diligently as the number of squares following that square increments. Taking into account that in the blockchain-based methodology all vitality exchanges are copied and shared over the system hubs, it is basic to give strong methods for ensuring this information. Actually, all the data put away as exchanges in the conveyed record are open, subsequently, to guarantee protection, conservation, and new techniques; for example, zero-learning verifications are utilized permitting one gathering, named the verifier, to check if the other party, named the prover, has mystery data without the prover disclosing the data [15].

Simultaneously the private-open keys help manufacture a validation and approval system in the appropriated record. The yields of vitality exchanges are not sent straightforwardly to a location of the beneficiary yet rather to a content that contains the open key of the beneficiary. The yield content of any exchange contains a lot of guidelines that must be implemented at whatever point the vitality is executed again later on. Along these lines the advantages are bolted, and the accompanying exchange should give the required information created utilizing the private key through an opening content. A second hindrance for interest reaction to the board is that a lot of information must be shipped, put away, and handled to adequately control the DR, and this turns out to be considerably increasingly basic for ongoing frameworks. On account of unified administration draws near, the coordination and interoperability of vitality information gathered from heterogeneous and disseminated DEP is testing. In the disseminated record case, the information obtained from the IoT shrewd metering gadgets is put away locally in squares as exchanges and repeated for approval to peer hubs. Since the number of exchanges can be high, a lot of exchanges that happen in a brief time frame are assembled in a solitary square and encoded utilizing Merkle trees including hash pointers. This gives expanded execution and diminishes the length for the chain of squares and furthermore the size of the squares to be repeated. All vitality exchanges in a square are combined in pairs, and the Merkle tree is steadily worked from base to top utilizing the hashes of exchanges until the root is come to. The hash of the root encodes the whole accumulation of exchanges that are recorded and totaled in the square and can be utilized by the hubs that need more stockpiling abilities (i.e., light hubs). The light hubs store just the header of the squares, while the genuine exchanges are put away remotely. The Merkle tree root gives enough data to the light hubs to have the option to check the consistency of the chain and, given the correct contribution, to check the participation of various exchanges in the square. The light hub can examine other system hubs (i.e., full hubs) for extra data in the event that it needs to confirm if an exchange was mined and to recognize the square that stores the real exchange. Simultaneously, control and authorization systems might be executed through which the support of a recently sent DEP to the appropriated system for DR programs the executives can be acknowledged or dismissed. In this sense, imaginative arrangements offering a high level of namelessness could be created utilizing blockchain stages, for example, Quorum (i.e., permission execution of Ethereum supporting information protection).

Blockchain conveyed agreement is utilized for DR check and close continuous DR money-related repayment utilizing the data assembled from each DEP part of the framework on the portion of vitality adaptability that was really conveyed. These data are accumulated in blockchain squares, for all time enrolled, and repeated over the disseminated record. Since the information structures offered by blockchain depend on hash pointers, the subsequent advantage is that the whole record turns into a carefully designed log that can be adjusted uniquely by reregistering the hashes for all the accompanying squares, which is infeasible. In this manner, significant activity in the blockchain approach is that the system ought to all things considered concur on the substance of the record, which for our situation mirrors the vitality condition of the framework and the DR occasions effectively tended to. In our methodology, rather than having one specialist keeping all vitality exchanges unified, like the DSO, the obligation is similarly shared among each companion hub of the system. Each time new vitality exchanges are enlisted by a DEP, these qualities are checked by oneself authorizing shrewd contracts utilizing the DR occasion understandings and expected vitality adaptability levels. Since the brilliant contracts are sent in the system, each DR occasion guideline is implemented by each companion and approved in close constant over every one of the friends. Accordingly the choice on the genuine portion of contracted adaptability, which has been viably conveyed by each friend, and money-related repayment is collectively settled upon by the various system peers through agreement.

They by and large check the whole blockchain, and vitality adaptability exchanges are not completely "affirmed" until new squares are included. To accomplish accord between the hubs, a proof-of-work (PoW) convention has utilized Greedy Heaviest-Observed Subtree (GHOST) together with an application-specific integrated circuit (ASIC) safe hashing calculation (Dagger-Hashimoto) [36]. The PoW is vitality utilization in this way; it isn't extremely engaging when we talk about vitality effectiveness. In this way, we propose an elective form dependent on proof of stake (PoS). The PoS agreement can't be tried at this point since the PoS model will be accessible in Ethereum just toward the start of 1 year from now [37]. In the PoS the member stake is utilized to decide the probability of a system friend including the following square of vitality exchanges to the blockchain and mining and approving all vitality exchanges engaged with DR and the money-related repayment. PoS calculations for mining the following substantial square and approving related exchanges/benefits in the blockchain could be stretched out to the particular instance of DR with a perspective on giving expanded dependability of the DR projects and matrix activity. In each DEP, some portion of the lattice could play the job of vitality exchange validator and could be the digger of the following legitimate square. Each validator should claim some stake in the power arrange, for our situation the absolute remunerated DR impetuses got to date, which could be utilized as an assurance of the square's legitimacy. To maintain a strategic distance from an incorporated choice where just the most excessive part settles on the approval choice, some level of randomization ought to be given. Arrangements like the one given by PPCoin shared digital currency [38], which consolidates flip coin randomization with the coinage factor (for our situation the DR impetuses age), could be received. Because of the blockchain-based methodology, the legitimate vitality exchanges and the real portion of actuated vitality adaptability (i.e., vitality request deviations from benchmark) are known in close continuous, and new DR occasions can be produced to manage impromptu circumstances.

In short the PoS-distributed consensus-based DR program verification works as follows:

(1) The transactions are registered by each DEP and shared with all the other energy parties interested to the same share of flexibility (not only DSOs but also TSOs, retailers, etc.) to be validated and mined in future blocks.
(2) The blocks are replicated, and the distributed ledger is updated to reflect the state of the grid.
(3) For each DEP, self-enforcing smart contracts check if the share of energy flexibility actually provided matches the expected levels agreed in the DR events.
(4) All DEPs collectively verify the entire blockchain, and the energy transactions are not considered to be fully "confirmed" until they are validated and aggregated in new blocks that are added to the ledger.
(5) The share of contracted flexibility effectively delivered, and the financial settlement are calculated using ledger information and the reward/penalty rates defined in the DR events.

The blockchain-based management of the smart energy grids provides solutions to many of the problems identified for the traditional, centralized approach as depicted in Table 1.

The appropriation of the blockchain ideas will change the shrewd matrix into a blockchain-based network that never again depends on a focal specialist yet can take any choice through keen contract standards authorized and checked by each DEP of the framework. Besides the conventional brought together administration of the brilliant lattice that is inclined to single purpose of disappointment vulnerabilities is supplanted with a decentralized methodology, where the insights, exchanges, control administrations, and installment settlements are altogether processed and checked in an appropriated way by every hub in the system.

TABLE 1 Assessment between traditional method and blockchain method [14].

Matter	Current method	Blockchain method
Failure in single point	Yes	No
Energy outline obscurity	No	Yes
Payment scheme	Centralized	Peer-to-peer sales/purchase system
Payment clearance	By central consultant Up to 60 days	It should be the agreement between all nodes Near real time
Energy outlines integration and aggregation	By central consultant	Over distribution ledger and agreement between all nodes
Demand response layout	By central consultant	Autonomous signaling through node cooperation and smart contracts
Energy agreement confirmation	By central consultant	Over agreement among all nodes

6. Smart contracts for demand response programs

In this methodology a brilliant contract is a bit of code that characterizes the normal vitality adaptability levels of each DEP for support in DR programs and the standards for guaranteeing the framework level harmony between vitality request and vitality generation. The guidelines may portray the conduct of DEPs during interest reaction occasions or may even address different requirements for keeping up the lattice steadiness and unwavering quality. These agreements are enrolled in the blockchain and are activated by new vitality exchanges (i.e., enlisting new vitality information from the IoT shrewd meters), which will make each blockchain hub update its state dependent on the outcomes got in the wake of running the savvy contract. In any case, regardless of whether the term is "contract," the shrewd contract ought to be an operator that has state factors, implements the related standards, and can be activated any time after its fruitful organization.

The deliberate enrolment of each DEP in a DR occasion is managed to utilize self-authorizing shrewd contracts. Such an agreement characterizes the DEP's gauge vitality profile, currently observed vitality esteems, and expected vitality profile, incorporating the normal changes as far as the measure of vitality adaptability to be moved during DR occasion time interims (see Table 2).

TABLE 2 State variables of smart contract and guidelines of DEPs [14].

State variable	Explanation
Baseline energy outline	Regular energy outline of a DEP regulated as an average of past measured energy principles; replicates how much in the nonattendance condition of the DR the DEP would have been consumed
Current energy profile	By using IoT smart energy metering devices, time series of obtained values acquired
Commanded energy profile	The signal provided by the DSO is requested to regulate its energy outline to a convinced level during the DR happening period

Other than the brilliant contracts customized and connected with every individual DEP, we have characterized savvy gets that actualize the principles for adjusting the vitality condition of the whole network and implementing its dependability. At the end of the day, it characterizes the guidelines for following and amassing enrolled at the degree of each DEP with the general objective of coordinating and adjusting the general vitality creation and utilization at the lattice level. On account of distinguishing irregular characteristics among generation and utilization, the shrewd contract starts new DR occasions and imparts to the intrigued DEPs the related DR signal and related motivating force and punishment rates. Table 3 presents the state variables controlled by the smart contract.

TABLE 3 Smart agreement modifiable [14].

State variable		Description
Grid energy state		The balance among energy consumption and generation at the smart grid environment defined as a sum of individual imbalances followed at the level of each DEP
New DR arrangements	Commanded energy outlines for DEPs DR returns and penalty rates	New DR indications defined by the DSO for bringing the smart grid into a balanced energy condition The rate used to compute the encouragement presented as a recompense for ensuing a DR indicator. The penalty percentage imposed for nonfulfillment

7. Validation and results

To approve the interest reaction to the board decentralization through block-chain and shrewd contracts, a reproduction-based model was executed utilizing the Ethereum stage. The vitality utilization information is given as a contribution of the reproduction procedure considering the vitality profiles of various UK structures distributed by administrative offices. The blockchain put together circulated methodology centers with respect to the medium- and low-voltage matrix and is given by electric power conveyance structure and stress factors that influence the keen vitality lattice improvement and the board. Generally the high-voltage current is changed over to medium voltage and is done utilizing electrical cables to the end clients (e.g., medium voltage organizes arrangement at city level). Auxiliary transformers convert the voltage from the medium- to low-voltage level, appropriate for direct utilization by the end clients. From these transformers, low-voltage systems branch off to the client associations furnished with shrewd vitality metering gadgets. Notwithstanding this plan a few elements stress the vitality matrix activity: (1) expanded portion of renewables (sunlight-based photovoltaic, wind turbines) that makes more irregularity and unpredictability in the vitality supply, and (2) circulated age that makes the homes and little scale organizations little vitality makers (i.e., DEPs) that are associated with the nearby low/medium conveyance arrange. This can be tricky at low- and medium-voltage system levels when nearby dissemination organizes and metering frameworks can't suit turn around streams, or when there are high sustainable power source creation tops and insufficient vitality request to cover them. For instance, in Europe, the absolute introduced limit of photovoltaic frameworks came to 69 GW in 2012, 80% of which was associated with low-voltage systems. Simultaneously, Europe remains an overall chief as for conveyed limit of sun-oriented power, and with a viewpoint for development of over 80% by 2020, which will require novel interest reaction, the board arrangements as the conventional ones won't most likely scale with the expanding number of DEPs. In specific areas in Italy, 20% of disseminated creation is sustained into the conveyance organize. In this circumstance, without a coordinating interest, the appropriation substations battle to effectively oversee invert streams (and guarantee generally speaking matrix soundness). In Germany, sunlight based and wind age must be disengaged from the lattice on occasion in light of the fact that sustainable power sources created a degree of intensity that the framework couldn't oblige. In Belgium the power framework experienced difficulty obliging the creation of sustainable power sources on bright and blustery days when there was very little modern interest. UK DR and adaptability market is evaluated to be just a 10th of its potential size and is attempting to scale quick enough. In this specific situation, our blockchain-based methodology can give new arrangements prepared to drive the change from customary market approaches and brilliant matrix tasks into novel decentralized and network-driven vitality frameworks. Simultaneously the appropriation of blockchain innovation for matrix the executives could be the beginning stage for empowering multipartner showcases just as market progression of region organize activity where vitality aggregators will join little scale vitality adaptability of DEPs and retailers will be in charge of the market supply of the past power product to the last clients. The European interest reaction the executive's framework market is relied upon to develop from $1.35 billion of every 2014 to $6.37 billion out of 2019, including a development rate of 36.3% during the figure time frame. Market patterns for interest reaction completely relate with the developing innovation pattern, being driven by the development in discontinuous age, reception of brilliant vitality metering foundations, and very factor loads that increment the instability of free-market activity. The expanded portion of renewables and dissemination harmonizes with the move toward savvy vitality metering and electric accusing stations and of the squeezing need to build the productivity of conventional appropriation systems. On the interesting side, numerous components will add to expanded pinnacle power requests, for example, time, speed, and area of electric vehicle charging. Fulfilling the expansion in pinnacle power request brought about by vehicle charging will require generous venture if their charging is unconstrained or uncontrolled. This is especially the situation if neighborhoods with low-voltage power matrices require a high volume of vehicle charging stations, as this will require considerable expensive system fortifications. These improvements are prompting an adaptable and progressively decentralized interest/supply load the executive's portfolio, which brings various difficulties, however in the long haul could give an answer to oversee pinnacle request and framework adjusting. The outcome isn't just market value instability or emotional swings in net burden yet, in addition, a decrease in an opportunity to adjust the electric framework. An opportunity to settle on basic working choices dependent on fast changes in the supply and interest is diminishing from moment interims to second interims. These shorter timescales make critical difficulties for powerful human association in the present choice procedures and by and large operational framework. This incorporates the brief length reactions required to keep up conveyance unwavering quality, control quality, and related operational variables, which isn't practical any longer utilizing brought together houses based DR the executives draw near. The execution of shrewd self-implementing contracts can acquire the essential degree of decentralization DR occasions the executives, taking into account the ID in close ongoing critical deviations that may influence network steadiness and require request adjustment. In addition, the blockchain-based innovations have the capability of changing the DR market and the executives to a completely decentralized one in which individual DEPs will be accountable for controlling the DR program's satisfaction, while the budgetary settlement should be possible through agreement and approval all things considered.

8.   Conclusions

In this section, we propose a decentralized framework for overseeing request reaction programs with regard to smart grids. We coordinate the components of the lattice with blockchain design and related brilliant contracts to guarantee the automatic meaning of expected vitality adaptability levels, the approval of DR understandings, and harmony between vitality request and vitality creation. A model was executed in Ethereum to approve and test the blockchain-based decentralized administration utilizing vitality hints of UK building datasets. The outcomes are promising, demonstrating that the matrix is prepared to do opportune changes of the vitality request in close continuous by establishing the normal vitality adaptability levels and approving all the DR understandings. Additionally, it makes ready for setting up an unadulterated distributed decentralized vitality exchanging instrument, which we exclude any go-between outsider like the DSO, with an effect in terms of vitality exchange cost decrease. In this manner, future enhancements will intend to actualize multipartner markets (DSOs, TSOs, retailers as contenders, or cooperators for similar vitality adaptability) utilizing our blockchain-shared interest reaction to the executive stage.

References

[1] Y. Li, W. Yang, P. He, C. Chen, X. Wang, Design and management of a distributed hybrid energy system through smart contracts and blockchain, Appl. Energy 248 (2019) 390–405.

[2] A. Meeuw, S. Schopfer, F. Wortmann, Experimental bandwidth benchmarking for P2P markets in blockchain managed microgrids, Energy Procedia 159 (2019) 370–375.

[3] X. Wang, W. Yang, S. Noor, C. Chen, M. Guo, K.H. van Dam, Blockchain-based smart contract for energy demand management, Energy Procedia 158 (2019) 2719–2724.

[4] Z. Li, S. Bahramirad, A. Paaso, M. Yan, M. Shahidehpour, Blockchain for decentralized transactive energy management system in networked microgrids, Electr. J. 32 (4) (2019) 58–72.

[5] X. Huang, Y. Zhang, D. Li, L. Han, An optimal scheduling algorithm for hybrid EV charging scenario using consortium blockchains, Future Generation Comput. Syst. 91 (2019) 555–562.

[6] Z. Su, Y. Wang, Q. Xu, M. Fei, Y.C. Tian, N. Zhang, A secure charging scheme for electric vehicles with smart communities in energy blockchain, IEEE Internet Things J. 6 (2018) 4601–4613.

[7] Y. Wang, Z. Su, N. Zhang, BSIS: blockchain based secure incentive scheme for energy delivery in vehicular energy network, IEEE Trans. Ind. Inform. 15 (2019) 3620–3631.

[8] C. Liu, K.K. Chai, X. Zhang, E.T. Lau, Y. Chen, Adaptive blockchain-based electric vehicle participation scheme in smart grid platform, IEEE Access 6 (2018) 25657–25665.

[9] M. Stephant, K. Hassam-Ouari, D. Abbes, A. Labrunie, B. Robyns, A survey on energy management and blockchain for collective self-consumption, in: 2018 7th International Conference on Systems and Control (ICSC), October, IEEE, 2018, , pp. 237–243.

[10] X. Yang, G. Wang, H. He, J. Lu, Y. Zhang, Automated demand response framework in ELNs: decentralized scheduling and smart contract, IEEE Trans. Syst. Man Cybern. Syst. 50 (2019) 58–72.

[11] J. Kang, R. Yu, X. Huang, S. Maharjan, Y. Zhang, E. Hossain, Enabling localized peer-to-peer electricity trading among plug-in hybrid electric vehicles using consortium blockchains, IEEE Trans. Ind. Inform. 13 (6) (2017) 3154–3164.

[12] X. Wu, B. Duan, Y. Yan, Y. Zhong, M2m blockchain: The case of demand side management of smart grid, in: 2017 IEEE 23rd International Conference on Parallel and Distributed Systems (ICPADS), December, IEEE, 2017, , pp. 810–813.

[13] S. Noor, W. Yang, M. Guo, K.H. van Dam, X. Wang, Energy demand side management within micro-grid networks enhanced by blockchain, Appl. Energy 228 (2018) 1385–1398.

[14] C. Pop, T. Cioara, M. Antal, I. Anghel, I. Salomie, M. Bertoncini, Blockchain based decentralized management of demand response programs in smart energy grids, Sensors 18 (2018) 162.

[15] A. Jindal, G.S. Aujla, N. Kumar, SURVIVOR: a blockchain based edge-as-a-service framework for secure energy trading in SDN-enabled vehicle-to-grid environment, Comput. Netw. 153 (2019) 36–48.

[16] C. Pop, T. Cioara, M. Antal, I. Anghel, I. Salomie, M. Bertoncini, Blockchain based decentralized management of demand response programs in smart energy grids, Sensors 18 (1) (2018) 162.

[17] V. Grewal-Carr, S. Marshall, Blockchain enigma paradox opportunity [online], Available at: https://www2.deloitte.com/content/dam/Deloitte/uk/Documents/Innovation/deloitteuk-blockchain-fullreport.pdf, 2016.

[18] J. Mattila, Industrial Blockchain Platforms: An exercise in use case development in the energy industry [online], Available at: https://www.etla.fi/wp-content/uploads/ETLAWorking-Papers-43.pdf, 2016.

[19] A. Baliga, Understanding blockchain consensus models [online], Available at: https://pdfs.semanticscholar.org/da8a/37b10bc1521a4d3de925d7ebc44bb606d740.pdf, 2017.

[20] Z. Zibin, An overview of blockchain Technology: Architecture, Consensus, and future trends [online], Available at: https://www.researchgate.net/publication/318131748_An_Overview_of_Blockchain_Technology_Architecture_Consensus_and_Future_Trends, 2017.

[21] M. Blinder, Making Cryptocurrency more environmentally sustainable [online], Available at: https://hbr.org/2018/11/making-cryptocurrency-moreenvironmentally-sustainable, 2018.

[22] C. Kim, Take Two: Ethereum is getting ready for the Constantinople hard fork redo [online], Available at: https://www.coindesk.com/take-two-ethereum-isgetting-ready-for-the-constantinople-hard-fork-redo, 2019.

[23] J. Bentke, Proof of Authority [online], Available at: https://energyweb.atlassian.net/wiki/spaces/EWF/pages/11993089/Proof+of+Authority, 2018.

[24] PwC global power & utilities, Blockchain – an opportunity for energy producers and consumers, https://www.pwc.com/gx/en/industries/assets/pwc-blockchain-opportunity-for-energy-producers-and-consumers.pdf, 2016. (Accessed 22 November 2017).

[25] Power Ledger - Where Power meets Blockchain, Power Ledger – A New Decentralised Energy Marketplace [Online], Available: https://powerledger.io/. (Accessed 1 May 2019).

[26] Multi apartment PV accounting, Multi apartment PV accounting [online], Available: http://www.key2.energy/. (Accessed 1 May 2019).

[27] Transactive Grid, LO3 Energy [online], Available: http://lo3energy.com/transactive-grid/. (Accessed 1 May 2019).

[28] "DAJIE!, [Online]. Available: https://www.dajie.eu. (Accessed 1 May 2019).

[29] Share&Charge, FAQs – easily charge your electric car with reliable charging station providers [Online], Available: https://shareandcharge.com/en/faqs-2/. (Accessed 1 May 2019).

[30] NRGcoin—Smart Contract for green energy, NRGcoin — Smart Contract for green energy [Online], Available: http://nrgcoin.org/faq. (Accessed 1 May 2019).

[31] EA.NRW, SEV Grünstromjetons: Blockchain-Anwendung macht Grünstromverbrauch sichtbar [Online], Available: http://www.energieagentur.nrw/eanrw/sevgruenstromjetonsblockchainan wendungmachtgruenstromverbrauchsichtbar. (Accessed May 2019).

[32] SolarCoin, FAQs [online], Available: https://solarcoin.org/en/faqfrequently-asked-questions/. (Accessed May 2019).

[33] The Sun Exchange, The Silicon Based Economy - financing solar cells with Bitcoin, 29 January 2017. [online]. Available: https://thesunexchange.com/silicon-based-economy-financing-solarcells-bitcoin. (Accessed May 2019).

[34] Metering.com, Bankymoon launches blockchain smart meter technology, [online]. Available: https://www.metering.com/bankymoonlaunches-bitcoin-to-simplify-utility-revenue-collection/ (Accessed May 2019).

[35] Dagger-Hashimoto, Available online: https://github.com/ethereum/wiki/blob/master/Dagger-Hashimoto.md. (Accessed 20 September 2017).

[36] Trustnodes.com, Ethereum Upgrade to Proof of Stake in 2018, Available online: http://www. trustnodes.com/2017/09/25/ethereum-may-upgrade-proof-stake-2018-says-vitalik-buterin. (Accessed 20 September 2017).

[37] S. King, S. Nadal, PPCoin: Peer-to-Peer Crypto-Currency with Proof-of-Stake, Available online: https://peercoin.net/assets/paper/peercoin-paper.pdf. (Accessed 20 September 2017).

[38] F.P. Sioshansi, Smart Grid: Integrating Renewable, Distributed and Efficient Energy, first ed., Academic Press, Cambridge, MA, USA, ISBN: 978-0-123864-53-6, 2011.

Further reading

The Value of the Engaged Energy Consumer, Quantifying the Value of Strong Customer Relationships for European Utilities, Available online: http://energypost.eu/wp-content/uploads/2014/ 12/COM-WP_Value-CE-EMEA-141017-PRINT-2.pdf. (Accessed 2 January 2018).

D. Leach, P. Henville, R. Wyatt, N. Hewitt, I. Morrissey, J. Weller, I. Tindall, N. Bachiller-Jareno, Continuous Measurements of Meteorological Parameters, Available online: http://nora.nerc.ac. uk/id/eprint/512741/. (Accessed 18 July 2017).

Chapter 9

Blockchain for decentralized optimization of energy sources: EV charging coordination via blockchain-based charging power quota trading

Sijie Chen[a], Jian Ping[a], Zheng Yan[a] and Wei Wei[b]
[a]Department of Electrical Engineering, Shanghai Jiaotong University, Shanghai, China,
[b]Department of Electrical Engineering, Tsinghua University, Beijing, China

1. Introduction

1.1 Motivation

Increasing electric vehicle (EV) penetration brings potential challenges to the secure operation of a distribution network. One of challenges is that simultaneous EV charging may result in overloads of distribution-side facilities [1]. Hence it is a key issue to design an effective EV charging coordination mechanism. To this end, an EV charging coordinator is necessary to enable this, but unfortunately in practice, such a coordinator may not exist. This study aims to address these two issues, that is, how to meet individual EV charging preferences while considering security constraints, and how to achieve this given the absence of a central coordinator.

1.2 First contribution

There are remarkable studies on coordinating EV charging stations. In some researches, a central operator collects charging demands of all charging stations (or all EVs) and optimizes the charging schedules while considering system constraints [1–5]. However, centralized optimization becomes impractical with a huge number of integrated EVs. To release the computational burden of the central operator, decentralized methods are introduced to the charging coordination problem [6–8]. The optimal charging schedule is achieved through

Blockchain-based Smart Grids. https://doi.org/10.1016/B978-0-12-817862-1.00009-9

iterations between the central coordinator and all charging stations (or all EVs). Though these methods significantly reduce the computational cost of the coordinator, frequent communication between the coordinator and charging stations is needed, which may be impractical.

Our first contribution is to propose a two-stage EV charging coordination mechanism. The permissible charging power of a charging station is represented by charging power quotas. At the first stage, the charging power quotas are initially allocated in an equitable and secure manner. At the second stage, the charging power quota trading is enabled via a double auction. Charging stations with elastic demand can sell charging power quotas to charging stations with inelastic demand. This yields the Pareto optimal allocation of charging power quotas.

1.3 Second contribution

To organize the charging power quota trading among charging stations, a reliable coordinator is indispensable. Recently, blockchain technology is introduced to the power grid [9, 10]. On blockchain, the energy trading can be organized without a central coordinator. The fairness and correctness of the trading is guaranteed by cryptography, ensuring trustful, decentralized, and transparent operation. Hence, blockchain is fit for the coordinator of charging stations. Study of Kang et al. [11] enables local P2P electricity trading between EVs in a charging station to maximize the total welfare of all parking EVs. In [12], the demand fluctuation level is reduced by running an iceberg order algorithm on blockchain. Study of Su et al. [13] satisfies EVs' individual needs by designing a smart contract. However, in these studies, the security of distribution facilities is rarely considered.

Our second contribution is to implement the proposed coordination mechanism on blockchain. The charging power quota is a typical digital asset, which is suitable for trading and settlement on blockchain. A decentralized, trustful, and transparent double-auction-based charging power quota trading platform is enabled by designing a smart contract. Simulation on an Ethereum private chain verifies the effectiveness of the proposed mechanism and our trading platform on blockchain.

2. EV charging power quota transaction mechanism

In this chapter, the EV charging schedules are coordinated in a two-layer structure. The lower layer is the EV charging station layer. A charging station collects the battery status and the charging demands of EVs, and charges the integrated EVs according to the allocated charging power quota. The upper layer is the coordinator layer. The charging power quotas are coordinated and allocated to charging stations to maximize overall welfare.

The proposed EV charging power quota transaction mechanism features two stages:

1. At the first stage, the distribution system operator (DSO) forecasts the conventional load and calculates the capacity margin of facilities, for example,

the transformer at the T-D station. At the same time, each charging station submits its charging demand. If the summation of the charging demand of all charging stations does not exceed the capacity margin, all demands can be met. Otherwise, the capacity margin is distributed to charging stations according to the proportion of the submitted demand of each charging station. At this stage, the charging power quotas are allocated equitably whereas the difference of charging urgency across charging stations is ignored.

2. At the second stage, charging stations can trade their charging power quotas with each other through a double auction. Charging stations with inelastic demand have strong willingness of charging, so they are the buyers in the double auction market. In contrast, charging stations with elastic demand are sellers in the market. By matching the submitted bids and offers from charging stations, buyers meet the inelastic charging demand, whereas sellers receive profits by selling charging power quotas. This yields Pareto improvement.

We divide a day into 48 periods. In each period, the charging power quotas in the next period will be allocated and traded. For example, energy delivered between 18:30 and 19:00 hours is traded during 18:00–18:30 hours and is detailed as follows.

Step 1: At 18:00 hours, the DSO calculates and broadcasts the capacity margin of the transformer at the T-D station during 18:30 and 19:00 hours according to conventional load forecast.

Step 2: Between 18:00 and 18:05 hours, all charging stations submit their charging demand during 18:30–19:00 hours by collecting the charging demands of on-site EVs.

Step 3: Between 18:05 and 18:07 hours, a charging demand satisfaction rate β is calculated as

$$\beta = \begin{cases} 1, & \sum_{i \in \Omega_E} P_i \leq P_M \\ \\ P_M \Big/ \sum_{i \in \Omega_E} P_i, & \sum_{i \in \Omega_E} P_i > P_M \end{cases} \tag{1}$$

where P_M is the capacity margin of the transformer in the next period broadcast by the DSO, Ω_E is the set of charging stations, and P_i is the charging demand of station i.

The initial charging power quotas of stations are determined by β. If the total charging demand is less than the capacity margin, the allocated charging power quota of each station during 18:30–19:00 hours is equal to its submitted demand and the following steps are skipped. Otherwise, to avoid the overload of the transformer, the initial charging power quota of each station is calculated as

$$P_i^{init} = \beta P_i \tag{2}$$

where P_i^{init} is the initial charging power quota distributed to station i.

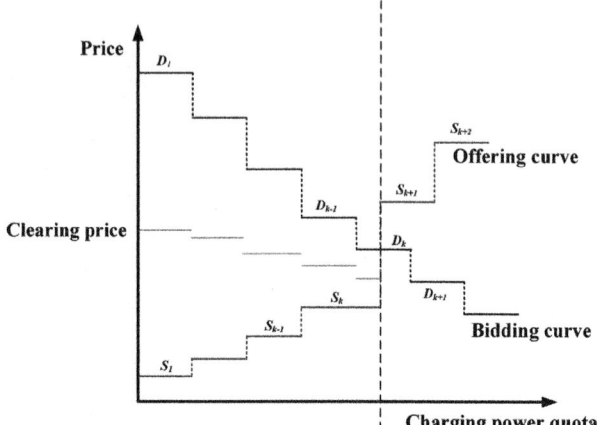

FIG. 1 Double auction mechanism of the charging power quota.

Step 4: After all charging stations get the initial charging power quotas, at 18:07–18:10 hours, all stations can evaluate the elasticity of the charging demand and then submit the quantity and the price of charging quota they are willing to purchase from/selling to other stations.

Step 5: Between 18:10 and 18:15 hours, bids and offers from charging stations are cleared using double auction mechanism. The mechanism is illustrated in Fig. 1. First, the mechanism orders the bids/offers from buyers/sellers. Second, the mechanism matches bids and offers orderly until the highest bid is lower than the lowest offer. Clearing prices are the average of bids and offers as illustrated in Fig. 1. Bids and offers on the left side of the dashed line are cleared. After the charging power quota trading market is cleared, all stations will be settled and their charging power quotas will be updated according to the transaction results.

Step 6: Between 18:30 and 19:00 hours, charging stations charge the on-site EVs according to the final charging power quotas. And all charging stations pay the charging fees to the DSO. The actual charging load is monitored by smart meters and sent to the DSO. If the actual charging load is higher than the charging power quota, the station will be penalized because it disturbs the secure and stable operation of the distribution network.

3. Implementation via Ethereum

To build a trustful, transparent, and decentralized trading platform, we implement the proposed mechanism on Ethereum by designing a corresponding smart contract. The coordination processes are divided into three procedures: charging demand submission, charging power quota double auction, and settlement. Fig. 2 provides a flowchart of the charging power quota trading on Ethereum. The details of the associated functions are explained as follows.

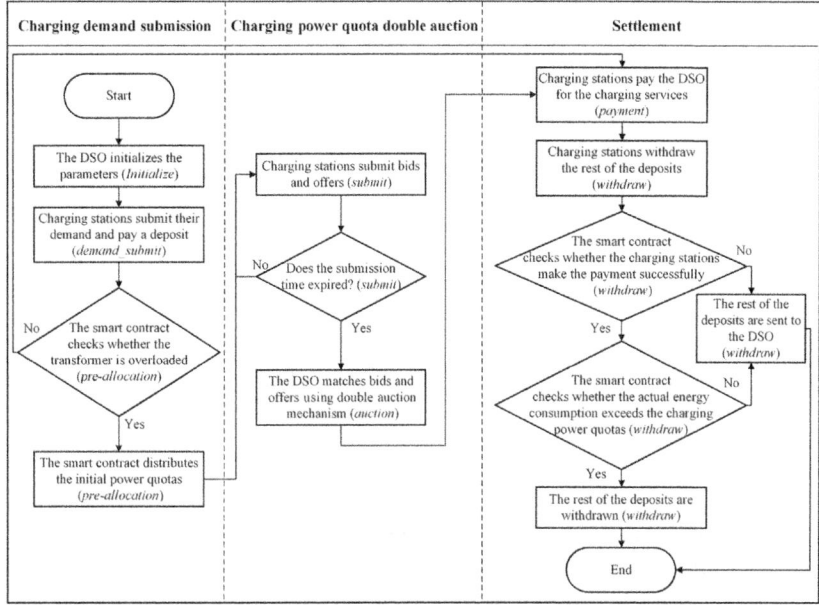

FIG. 2 Flowchart of charging power quota trading on Ethereum.

3.1 Charging demand submission

(1) *Initialize*

The *Initialize* function, refers to step 1 in Section 2, can be only called by the DSO at the beginning of the trading period. With this function, the currency exchange rate for Ether to the Token in the smart contract, the real-time electricity price, and the capacity margin of the transformer are updated. Due to the fluctuation of Ether's value, we choose the ERC-20 Token as the payment currency. By modifying the currency exchange rate for Ether to the Token, a Token is always worth 0.01 CNY. In the rest of the section, the Token is the unit of the price.

(2) *Demand_submit*

The *demand_submit* function refers to step 2 in Section 2. By calling this function before the deadline of the charging demand submission procedure, charging stations submit the charging demands for the next period to the smart contract. When calling this function, stations must pay deposits to the contract.

The deposits paid by charging stations have the following purposes. First, part of deposits will be used in the double auction market for the settlement of charging power quotas. Second, deposits contain the charging fees that will be paid to the DSO. And third, a penalty will be deducted from deposits if the actual power consumption of charging stations is more than their charging power quotas.

(3) *Pre-allocation*

The *pre-allocation* function refers to step 3 in Section 2. By executing this function, the smart contract distributes the initial charging power quotas by Eqs. (1), (2). If the total charging demand does not exceed the capacity margin of the transformer, procedure 2 will be skipped. Otherwise, the smart contract will trigger the double auction market. By monitoring the contract, charging stations can view their initial charging power quotas and the time when the double auction starts.

3.2 Charging power quota double auction

(1) *Submit*

The *submit* function refers to step 4 in Section 2. When the double auction market is triggered, charging stations can submit the purchase/selling prices of charging power quotas by calling this function. The smart contract will sort bids/offers in a descending/ascending order.

(2) *Auction*

The *auction* function refers to step 5 in Section 2. After the deadline of submission, the bids and offers are matched in the double auction market after the DSO calls this function. The details of the algorithm are shown in Algorithm 1.

Algorithm 1 Charging power quota double auction

1: Sort of bids in a descending order:
2: $bids = \{(bid_1, demand_1), ..., (bid_n, demand_n)\}$
3: Sort of offers in an ascending order:
4: $offers = \{(offer_1, supply_1), ..., (offer_m, supply_m)\}$
5: **while** $bids! = \{\}$ & $offers! = \{\}$ & $bid_1 \geq offer_1$ **do**
6:    $price = (bid_1 + offer_1)/2$
7:    $amount = min(demand_1, supply_1)$
8:    transaction(from: $buyer_1$, to: $seller_1$, value: $price * amount$)
9:    $(bid_1, demand_1) \leftarrow (bid_1, demand_1 - amount)$
10:    $(offer_1, supply_1) \leftarrow (offer_1, supply_1 - amount)$
11:    **if** $demand_1 == 0$ **then**
12:      **for** $(bid_i, demand_i) \in bids$ **do**
13:        $(bid_i, demand_i) \leftarrow (bid_{i+1}, demand_{i+1})$
14:      **end for**
15:    **else**
16:      **for** $(offer_j, supply_j) \in offers$ **do**
17:        $(offer_j, supply_j) \leftarrow (offer_{j+1}, supply_{j+1})$
18:      **end for**
19:    **end if**
20: **end while**

3.3   Settlement

(1) *Payment*

After energy delivery, all charging stations must pay the DSO for the charging services by calling this function. The values of the payments depend on the predetermined tariff and the actual energy consumption measured by smart meters.

(2) *Withdraw*

With this function, charging stations that have already called the *payment* function can withdraw the rest of the deposits. Moreover, the deposits will compensate the DSO if the actual energy consumption of charging stations is higher than their charging power quotas. This is also a penalty for the station in default. This function together with the *payment* function constitutes step 5 in Section 2.

4.   Case studies

4.1   Data

A case study is done on a distribution network under a 10-kV transformer in Shanghai. It is assumed that the capacity margin of the transformer is 343 kW. The real-time electricity prices are presented in Table 1. We simulate the charging power quota trading during 18:00–18:30 hours on an Ethereum private chain. A total of eight charging stations participated in the charging power quota trading, named as A–H. The charging demands are simulated by Monte-Carlo method.

4.2   Simulation results

In the charging demand submission procedure, first, the DSO calls the *initialize* function and charging stations call the *demand_submit* function. Then, by executing the *pre-allocation* functions, the allocation scheme of initial charging

TABLE 1 Real-time electricity prices.

Time (hours)	Electricity price (RMB/kWh)	Electricity price (token/kWh)
22:00–6:00	0.32	32
8:00–11:00 18:00–22:00	1.12	112
6:00–8:00 11:00–18:00	0.69	69

TABLE 2 Allocation scheme of initial charging power quotas.

Charging station	Charging demand (kW)	Initial charging power quota (kW)
A	36	29.82
B	48	39.75
C	42	34.78
D	66	54.66
E	30	24.85
F	66	54.66
G	54	44.72
H	48	39.75

power quotas is as presented in Table 2. The total charging demand of all charging stations is 390 kW. Therefore, the charging demand satisfaction rate $\beta = 0.83$.

In the charging power quota double auction market, the submitted bids and offers of charging stations are as presented in Table 3. After the DSO calls the *auction* function, all participants can view the result of the double auction market from the blockchain, as illustrated in Fig. 3.

TABLE 3 Bids and offers of charging stations in the double auction market.

Charging station	Demand (kWh)	Supply (kWh)	Unit price (token/kWh)
A	/	3.2	25
B	6.5	/	24
C	/	9.8	22
D	10.9	/	26
E	3.2	/	28
F	/	7.2	19
G	5.4	/	20
H	/	6.8	16

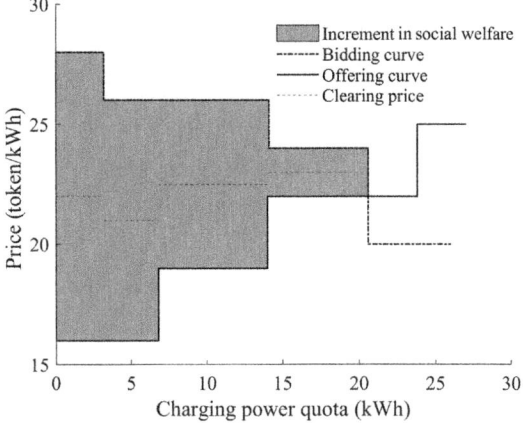

FIG. 3   Result of the charging power quota trading.

TABLE 4 Settlement of the charging stations.

Charging station	Charging service fee (token)	Payment in double auction (token)
A	3339.32	0
B	4452.43	73.6
C	3895.88	−225.4
D	6122.09	250.7
E	2782.77	73.6
F	6122.09	−165.6
G	5008.98	0
H	4452.43	−156.4

In the settlement procedure, all the charging stations pay the DSO and withdraw deposits. As a result, the changes in balances of charging stations are as presented in Table 4, where the minus payment represents an income.

The simulation results show that Ethereum can effectively coordinate charging stations. By interacting with the smart contract deployed on the Ethereum private chain, charging stations receive charging power quotas in a transparent and decentralized way. As a digital asset, charging power quotas can be conveniently settled on Ethereum.

In the charging power quota trading simulation, except charging stations A and G, other charging stations either receive profits by selling charging power quotas or satisfy their inelastic charging demand. Therefore, with the proposed mechanism, all participants can make Pareto improvement while the system security constraint is satisfied.

5. Conclusion

This chapter proposes a two-stage EV charging coordination method. In the proposed method, the charging power is allocated to charging stations in a secure, fair, and efficient way. The following charging power quota trading method enables charging stations to make Pareto improvement. Given the lack of a central coordinator, the proposed mechanism is implemented via the Ethereum blockchain. Transparency and effectiveness of the coordination are guaranteed by blockchain technology.

The simulation results show the effectiveness of the proposed coordination method. One can also see that the designed platform on Ethereum can efficiently coordinate charging stations, providing a practical solution with the absence of central coordinators in practice.

References

[1] B. Sun, Z. Huang, X. Tan, D.H.K. Tsang, Optimal scheduling for electric vehicle charging with discrete charging levels in distribution grid, IEEE Trans. Smart Grid 9 (2) (2016) 1, https://doi.org/10.1109/TSG.2016.2558585.

[2] Z. Li, Q. Guo, H. Sun, S. Xin, J. Wang, A new real-time smart-charging method considering expected electric vehicle fleet connections, IEEE Trans. Power Syst. 29 (6) (2014) 3114–3115, https://doi.org/10.1109/TPWRS.2014.2311954.

[3] L. Hua, J. Wang, C. Zhou, Adaptive electric vehicle charging coordination on distribution network, IEEE Trans. Smart Grid 5 (6) (2014) 2666–2675, https://doi.org/10.1049/ep.1982.0374.

[4] C. Sabillon Antunez, J.F. Franco, M.J. Rider, R. Romero, A new methodology for the optimal charging coordination of electric vehicles considering vehicle-to-grid technology, IEEE Trans. Sustain. Energy 7 (2) (2016) 596–607, https://doi.org/10.1109/TSTE.2015.2505502.

[5] Z. Liu, Q. Wu, S.S. Oren, S. Huang, R. Li, L. Cheng, Distribution locational marginal pricing for optimal electric vehicle charging through chance constrained mixed-integer programming, IEEE Trans. Smart Grid 9 (2) (2018) 644–654, https://doi.org/10.1109/TSG.2016.2559579.

[6] C.-K. Wen, J.-C. Chen, J.-H. Teng, S. Member, Decentralized plug-in electric vehicle charging selection algorithm in power systems, IEEE Trans. Smart Grid 3 (4) (2012) 1779–1789, https://doi.org/10.1109/TSG.2012.2217761.

[7] C. Shao, X. Wang, M. Shahidehpour, X. Wang, B. Wang, Partial decomposition for distributed electric vehicle charging control considering electric power grid congestion, IEEE Trans. Smart Grid 8 (1) (2017) 75–83, https://doi.org/10.1109/TSG.2016.2595494.

[8] T. Zhao, Z. Ding, Distributed initialization-free cost-optimal charging control of plug-in electric vehicles for demand management, IEEE Trans. Ind. Informatics 13 (6) (2017) 2791–2801, https://doi.org/10.1109/TII.2017.2685422.

[9] E. Mengelkamp, J. Gärttner, K. Rock, S. Kessler, L. Orsini, C. Weinhardt, Designing microgrid energy markets. A case study: the Brooklyn microgrid, Appl. Energy 210 (2017) 870–880, https://doi.org/10.1016/j.apenergy.2017.06.054.

[10] G. Zizzo, E. Riva Sanseverino, M.G. Ippolito, M.L. Di Silvestre, P. Gallo, A technical approach to P2P energy transactions in microgrids, IEEE Trans. Ind. Informatics, https://doi.org/10.1109/TII.2018.2806357.

[11] J. Kang, R. Yu, X. Huang, S. Maharjan, Y. Zhang, E. Hossain, Enabling localized peer-to-peer electricity trading among plug-in hybrid electric vehicles using consortium blockchains, IEEE Trans. Ind. Informatics 13 (6) (2017) 3154–3164, https://doi.org/10.1109/TII.2017.2709784.

[12] C. Liu, K.K. Chai, X. Zhang, E.T. Lau, Y. Chen, Adaptive blockchain-based electric vehicle participation scheme in smart grid platform, IEEE Access 6 (2018) 25657–25665, https://doi.org/10.1109/ACCESS.2018.2835309.

[13] Z. Su, Y. Wang, Q. Xu, M. Fei, Y.-C. Tian, N. Zhang, A secure charging scheme for electric vehicles with smart communities in energy blockchain, IEEE Internet Things J., https://doi.org/10.1109/JIOT.2018.2869297.

Chapter 10

Islanded microgrid management based on blockchain communication

Amin Shokri Gazafroudi[a], Yeray Mezquita[a], Miadreza Shafie-khah[b], Javier Prieto[a,c] and Juan Manuel Corchado[a,c,d,e]

[a]BISITE Research Group, University of Salamanca, Salamanca, Spain, [b]School of Technology and Innovations, University of Vaasa, Vaasa, Finland, [c]Air Institute, IoT Digital Innovation Hub (Spain), Salamanca, Spain, [d]Department of Electronics, Information and Communication, Faculty of Engineering, Osaka Institute of Technology, Osaka, Japan, [e]Pusat Komputeran dan Informatik, Universiti Malaysia Kelantan, Kota Bharu, Kelantan, Malaysia

1. Introduction

In the last decade, microgrids (MGs) have been introduced to decentralize power systems through the integration of distributed energy resources. The current energy supply network suffers from blackouts under the harsh conditions of natural disasters, such as those that occurred during the hurricanes Irma and Maria in the United States [1]. As a result, it is necessary to create a more reliable and resilient power grid. Blackouts can be prevented by dividing the distribution network into autonomous and self-managing subnetworks called MGs that are islanded from the main distribution network.

One of the characteristics of MGs is that they encourage renewable energy resources (RES), for example, photovoltaic panels in buildings and wind turbines in farms [2]. Also, they can be islanded from the main power grid, so they are more resilient to the natural disasters occurring in other zones [3]. Finally, they can be managed and optimized locally because they are smaller in size, resulting in lower energy costs.

1.1 Literature review

An MG consists of entities that are capable of producing energy (producers), consuming energy (consumers), or doing both at once (prosumers, e.g., batteries). To maintain a balance in the network while trying to use to the potential energy supplied by the RES, energy storage systems (ESS) consisting of batteries can be

Blockchain-based Smart Grids. https://doi.org/10.1016/B978-0-12-817862-1.00010-5

operated. The centralized control approaches for MG management have been proposed in refs. [4–6]. In these approaches a central controller is in charge of balancing the power in all the buses of the MG, adjusting the power flow of the distributed battery system according to the state of charge of the batteries.

Wu et al. [7] proposed a decentralized approach for the control and management of the MG. In Wu et al. [7] a hierarchical controller has been presented, composed of local controllers that coordinate the distributed elements of the MG without relying on external communication links. The MGs are managed locally, and the control of the MGs is distributed among the local controllers. In this way, each controller is defined as an agent. In a multiagent system (MAS), agents interact and negotiate with each other to achieve the system's global objective, to optimize the performance of the MG. This concept is also used to optimize the performance of platforms designed for different fields, such as precision farming [8] and control of chemical processes [9].

Examples of MAS architectures for distributed MG control have been proposed in refs. [10–12]. All these works are based on the hierarchical organization of three layers of controller agents. Usually the first layer consists of local controllers of the physical devices that are part of the MG. The second layer consists of an MG coordinator that coordinates the communication between the controllers of the MG. Finally the third layer consists of a controller that negotiates with other third layer agent controllers of different MGs to respond to the external demands of the agents presented in the MG.

To optimize the performance of the MG, agents negotiate with each other to reach agreements that minimize the expected costs. In Nunna and Doolla [11] a continuous double auction method is proposed to negotiate the distribution of the energy price among the platform. The buyer agents ask sellers about the electricity price. The strategy of the buyers requires information about the acceptable market price range for the auctioned energy. Also, they make use of the forecasted market price and the risk they can take. Eddy et al. [12] uses a similar approach to optimize the system.

Bui et al. [10] proposed an auction algorithm of sequential actions based on the double auction method. In this algorithm the agents cooperate instead of competing, performing a global optimization thanks to which maximum profit can be obtained. The producer agents share the local information related to the state of the charge of their batteries and the cost of their produced energy. The coordinator is in charge of balancing the network by distributing the energy stored based on the agents' demand and production at the minimum cost.

The resource optimization issues that an MG resolves along its automated responsiveness bring with it some concerns regarding the privacy of the data and security of the platform. The data transferred alongside the network can be sniffed and stolen by third parties that can take advantage of them. The security of the platform is also a problem; when there are no signature or encryption mechanisms, the platform is at risk of man-in-the-middle attacks, letting attackers send instructions to the platform's devices.

To solve the previously mentioned issues related to the use of smart devices in the MG, it has been proposed to integrate blockchain technology (BT) in those systems [13]. Some companies like the Pwr. Company [14], Powerledger [15], Key2Energy, Lo3 Energy—Transactive Grid and Brooklyn Microgrid [16], Dajie [17], Share&Charge [18], and NRGcoin [19] have used the BT to create autonomous energy markets in the MGs. In this way the BT improves the security of this kind of systems by providing a public/private key pair mechanism to sign and encrypt transmitted data between the actors of the platform. Another way of achieving this is by encrypting the data stored in the blockchain or by simply storing the hash of the data that is stored locally for each of its agents.

2. Blockchain

The importance of using the BT in MG systems lies in that they enable an automatic and autonomous peer-to-peer energy market inside the MG. Like the cryptomarket created with Bitcoin [20] or Ethereum [21], the use of the BT enables an energy market in which autonomous agents can negotiate and make transactions together with digitalized real assets, for example, energy and cryptocurrencies.

A blockchain is a distributed ledger in which data can be stored. The distributed database is denominated simply blockchain, while the BT is the technology that supports the database, for example, the network of nodes that keep a copy of the blockchain, the consensus mechanism that nodes use to add new data to the blockchain, and the initial configuration parameters of the network.

To add new data to the blockchain, a node must write it in a new block; it then will be concatenated to the previous block that is already stored in the blockchain. When a new block of data is created, it is broadcasted through the network of nodes to be validated by them. If they reach a consensus and the block is valid, it means it doesn't contain malicious data and each node adds it to its local copy of the blockchain. The way the node that is to create a new block is selected depends on the kind of consensus mechanism used by the network. The means by which a new node can begin its interaction with the blockchain defines the type of BT used. A blockchain is called public when a node can connect to the network and can read and write the data in the blockchain and participate in the validation process, all without the need for any permission. Public blockchains are the most popular type of blockchain, they are completely decentralized, and anyone can be part of the network.

If a node needs the permission of any kind from the network to be part of it, then the blockchain is permissioned. This kind of blockchains is characterized because there are defined roles for the nodes that participate in the network. This type of BT provides more privacy while improving the speed at which new data are added to the blockchain at the expense of having a more centralized platform.

There is a third type of BT where the network is owned by a unique institution. This kind of blockchains is called private blockchains and may be accessed by everyone outside the institution, but the consensus mechanism and the validation process depend on the institution that owns it. This characteristic makes the blockchain faster; the nodes are known, so 51% of attacks are prevented; and the privacy of the data stored in it is improved; however, it is a centralized solution.

2.1 Consensus mechanisms

The consensus mechanism is needed to get rid of some intermediaries and automate the process of storing the new data that had been validated previously. Thanks to the consensus mechanism, the nodes of the network reach an agreement when it comes to adding blocks to the blockchain. There are three most commonly used algorithms, the rest are their variations:

1. Proof of work (PoW): This algorithm solves a cryptographic problem through the addition of a new block to the blockchain. The effort the nodes of the network (miners) put into searching for that solution (work) is enough to prevent them from adding illegal transactions. The computational power required for the validation process is much lesser than the power required for mining; therefore sending spam becomes useless because the cost of malicious action is higher than the reward that can be obtained.
2. Proof of stake (PoS): Unlike in proof of work, in this consensus algorithm, the nodes collaborate with each other to find a solution for the addition of a new block, taking turns to add them. The node—that has to add the next block—is selected according to the amount of deposited coins (stake), assuming that it is going to be an honest node to prevent losing its escrow after the validation process.
3. Practical Byzantine fault tolerance (PBFT): In this consensus algorithm, each addition of a new block to the blockchain is called a round. In each round a node that proposes the new block is selected; for its validation the block must receive 2/3 of the votes of all the nodes in the network.

2.2 Smart contracts

The blocks are capable of storing different types of data, such as the quantity of assets in a transaction or the set of instructions of a program. The data stored in a blockchain are immutable and can be supervised in a distributed way [22]. This kind of scripts is called *smart contracts*, which facilitate, verify, or enforce a contract taking into account a set of predefined conditions [23]. They are self-executing and self-verifying contractual agreements that automate the life cycle of a contract to improve compliance, mitigate risk, and increase efficiencies between different parties [24].

Making use of smart contracts in the MGs allows for the tokenization of energy put on sale, the automatic transactions of energy between the platform agents, the audition of the real energy transactions from the seller to the buyer, and the storage of bids placed during the auction process of the internal energy market.

3. Platform architecture

In this section the proposed platform is going to be explained in detail. As shown in Fig. 1, the platform consists of a deployed MAS that controls the islanded MG; an MAS in charge of the data analytics of the entire system; a permissioned blockchain that acts as the "central" authority; and a hardware layer of devices for the production, storage, and measurement of the energy in the MG.

3.1 Proposed MAS

The MAS that controls the platform comprises intelligent agents that take decisions aimed at optimizing the system's global state; see Fig. 2. Some of the optimization actions taken by agents are buying energy at minimum costs when needed, storing enough energy to prevent voltage drops during consumption peaks, and continually supplying the energy requested by the network. The MAS is divided into different organizations that group agents according to their functionality:

1. MG operating system

   This part of the system is the one in charge of the proper functioning of the MG network. The agents grouped here monitor the state of the network and the demands of the entities written in the blockchain and the electricity price from the outside market while injecting energy to the islanded MG when it is needed, always maintaining the energy supply:

   a. MG operator agent (MGO)

   The MGO monitors the MG network to balance it and ensure its correct functioning. Other agents inform the MGO agent about the changes in the production and consumption of energy at every moment. The MGO receives the auditions made in the platform to make a decision against the audited entities.

   b. Blockchain event listener (BEL)

   The BEL agent monitors the blockchain and uses the data stored in it to read the state of the network. The BEL agent informs the MGO agent to act and maintain the balance when this agent detects a possible unbalance in the consumption and production of energy.

   c. Market crawler agent (MCA)

   The MCA is responsible for monitoring the electricity price of the wholesale market from the upstream grid. This agent uses the collected data provided by data analytics system to predict the electricity price and

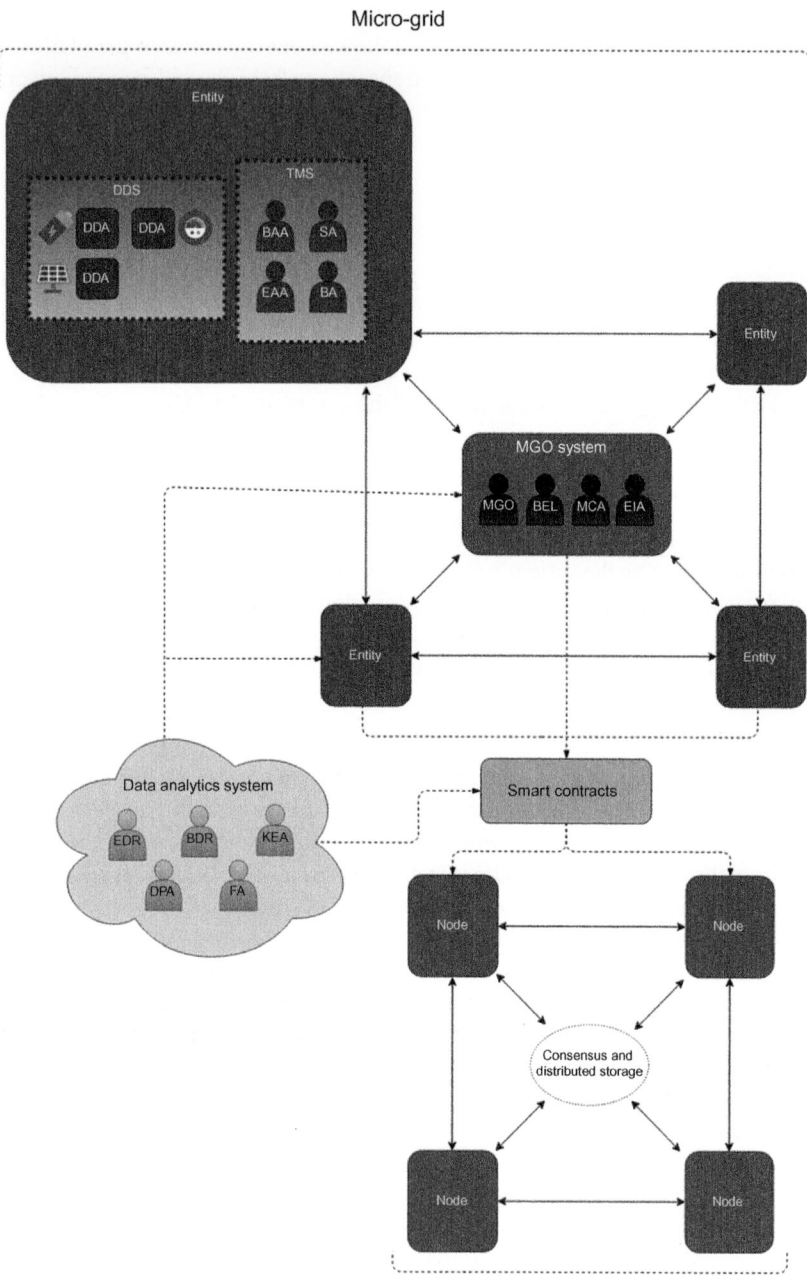

FIG. 1 Architecture of the deployed platform.

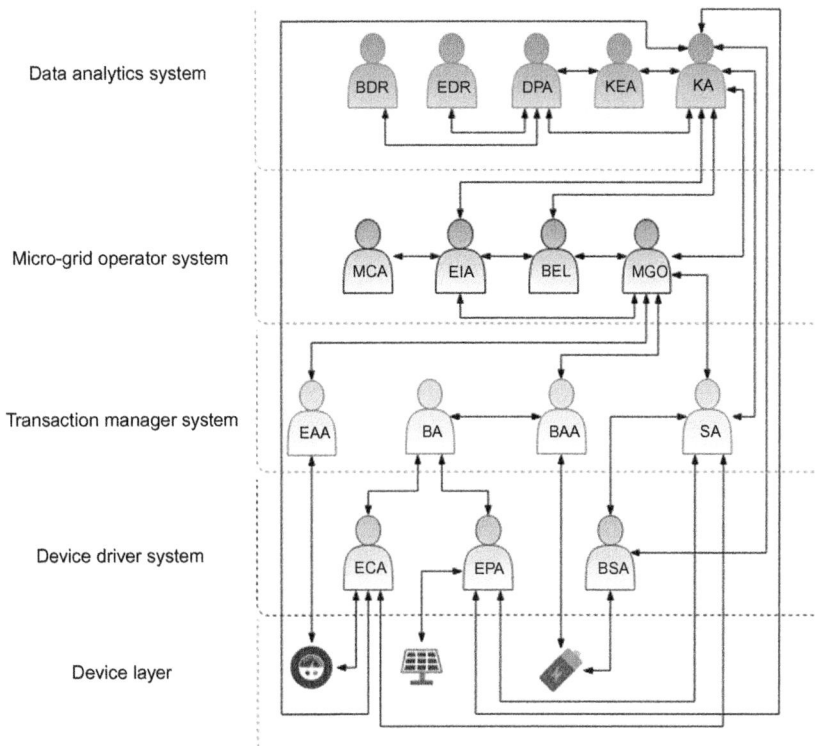

FIG. 2 Relation diagram of the agents that control the platform.

inform the energy injector agent (EIA) to make optimum decisions to buy energy.

d. Energy injector agent (EIA)

This EIA decides to buy energy from outside the MG island. It receives information about the state of the network from the MGO and the BEL agents and the wholesale market from the MCA agent. Based on the quantity of energy stored by the network, this agent makes decisions to acquire energy from upstream grid.

2. Data analytics system

The agents grouped in the data analytics system provide the entire platform with the ability to compile the data from inside and outside the blockchain and use those data for forecasting.

a. Blockchain data reader (BDR)

The BDR agent reads and normalizes data from the blockchain when it is needed by the data parser agent (DPA).

b. External data reader (EDR)

The EDR agent crawls and normalizes data from external sources when it is needed by the DPA.

c. Data parser agent (DPA)

This agent asks the data from the BDR and the EDR when the forecast agent (FA) or the knowledge extractor agent (KEA) requests it. Then it fuses, if necessary, and parses those data in an understandable way for the agents that request it.

d. Forecast agent (FA)

This agent responds with predictions to requests from other agents of the platform. To make those predictions the FA agent uses the models created by the KEA agent to make predictions with the data it obtains from the DPA agent.

e. Knowledge extractor agent (KEA)

The KEA agent is the one in charge of finding patterns in the data received by the DPA agent. To find those patterns, this agent uses unsupervised (clustering and dimensionality reduction) and supervised (support vector machine and neural network) machine learning algorithms to create models that are capable of making accurate predictions, for example, the quantity of energy that will be generated by the producers, and the fluctuations of the energy price of the external market.

3. Transaction manager system (TMS)

The TMS organizes the agents to manage energy flows among entities of the platform. The TMS is responsible for the tokenization of the sold energy, the management of the auction system that lets the consumers bid for lots of energy and the producers to sell their lots to the best bid, and the audition of the energy transmission between the network and the entities:

a. Buyer agent (BA)

The BAs are related to each entity of the MG that wants to consume energy. These agents search in the blockchain offers of the producer agents and bid for them to pay the minimum possible for the energy they need.

b. Battery audit agent (BAA)

The BAAs are autonomous and are related to batteries of the energy production stations. They are being created from code in the blockchain by the MGO, and they audit the amount of energy batteries that have tokenized energy. If the battery is part of a prosumer entity, the BAA will block the internal consumption of stored energy that is already tokenized.

c. Sale agent (SA)

The SAs are associated with each entity that can produce energy. These agents automatize the sale of energy to obtain the maximum benefit from the energy stored in their stations. It is also in charge of establishing the waiting time for getting the best bid. Besides, the SAs predict energy generation and the energy stored in the batteries associated with the entities.

With those predictions, they can know the maximum amount of energy they should put at the sale, the maximum waiting time for listening new bids, and the initial cost of the lots of energy put up for auction.

d. Energy audit agent (EAA)

The role of the EAAs is to audit the amount of energy that is transacted between the MG network and the entities within it. These are autonomous agents that store in the blockchain the log of all transactions of energy realized between points of the microgrid network. When the transaction between two entities does not correspond to the quantity of energy that is flowing to and from the entities, the smart contract this agent works informs the BEL agent through an event.

4. Device driver system

In this system the agents in charge of controlling the devices of the MG have been grouped together. The information collected by this system includes the entities' energy consumption and production and the state of the batteries that provide energy to the network. The agents are informed of the abnormalities through events in the blockchain:

a. Energy production agent (EPA)

The EPAs are in charge of writing in the blockchain the log of the energy production of a station for each sample time. The smart contracts these agents work with also inform the BEL agent when atypical measures are being taken of the stations they monitor. If an unusual measure is taken, it could mean that the production station is not functioning correctly and the data from this station cannot be used to make predictions.

b. Energy consumption agent (ECA)

The ECAs are responsible for writing in the blockchain the state of energy consumption of the entity they monitor. If the data show any sudden consumption peak, it could mean that the consumption is abnormally high, which possibly increases the error of the predictions and the risk of a voltage drop or blackout of the network.

c. Battery state agent (BSA)

The BSAs are associated with the batteries of the system and records in the blockchain the log of the energy stored at max capacity. When the max energy of a battery drops to a threshold, the BSA associated with it would inform that the battery needs maintenance.

3.2 Algorithm for the auction process

The auction process proposed begins when one of the entities of the platform makes an energy offer, as shown in Fig. 3. The offer is made only if the SA determines that surplus energy will be generated in the near future, on the basis of actual consumption data and forecast consumption (obtained from the ECA), actual energy production data and forecast energy production (obtained from

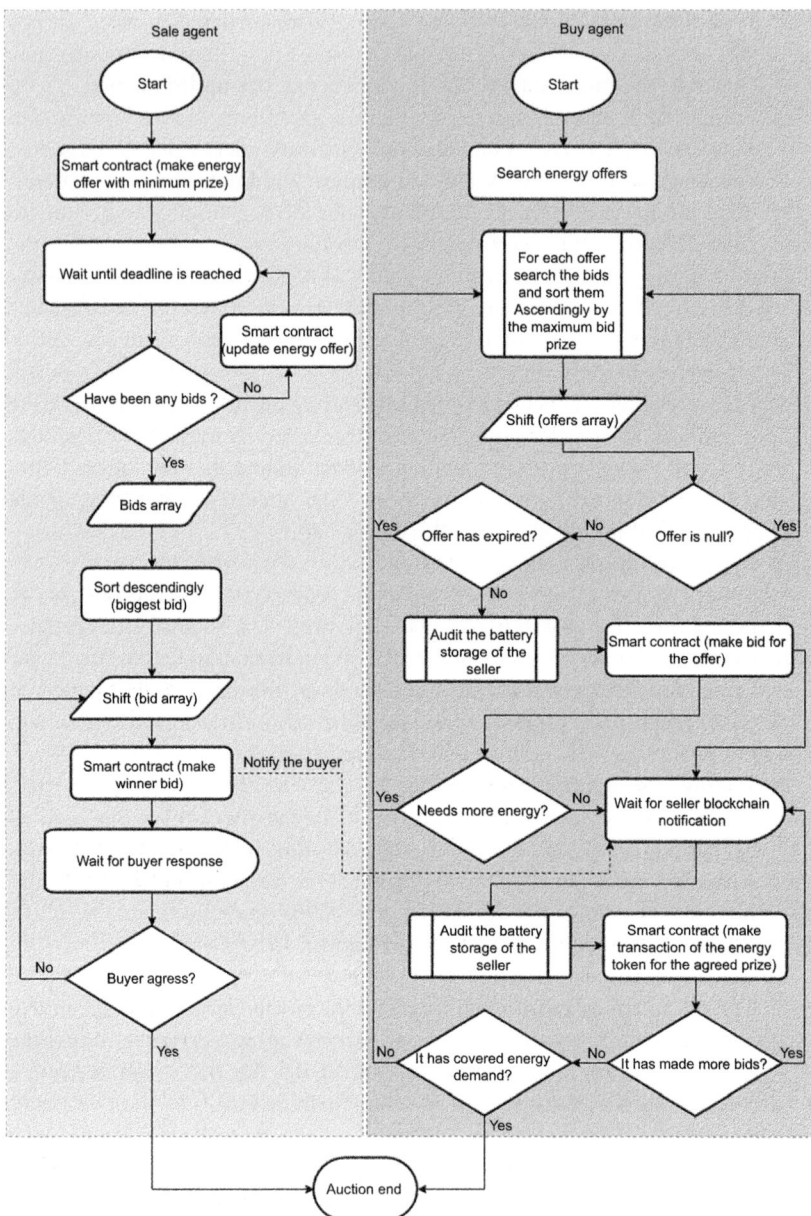

FIG. 3 Flow chart of the auction process.

the EPA and only if the SA is associated with a prosumer entity), and the energy stored in the monitored batteries (obtained from the BSA and only if the entity owns a battery). The SAs will offer all the surplus they consider appropriate; in the case of a prosumer entity, it has to take into account the quantity of energy it will need, assuming the energy it produces is cheaper than the energy it buys from others.

The energy offer of the MGO is always present and it doesn't expire, because it is in charge of maintaining the balance of the MG while optimizing the cost of the energy used in it. The first to bid for it is the one who obtains it automatically, starting the transaction of energy at that moment. The expiration of an offer can be used to let the SAs reduce the offer's initial price, just in case there aren't any bids for that offer.

BAs search the offers that are published in the blockchain when a consumer or prosumer entity knows that it will need energy in the near future. The BA obtains the amount of energy stored in its batteries from the BSA, the actual and forecast energy consumption from the ECA, and the actual and forecast production from the EPA, if applicable, to calculate the quantity of energy it will need.

After searching the offers published, for each one of them, a BA audits the stored energy of the batteries of the offer's entity. If the entity has enough stored power, the BA publishes the bid via a smart contract. If the BA needs to buy more energy, it proceeds to the next offer.

For each of the placed bids, the BA waits for the offer to expire or for a blockchain event that would inform that the bid has been exceeded by another. If a bid is being surpassed, the offer with the new price is added to the array of offers and placed in its corresponding order. Whenever a bid for an offer is published, the BA that has a selected bid by that offer is notified that its bid has been surpassed. When a BA is notified of this, it can place another bid or bids for a cheaper offer.

If the offer has received any bids when it reaches its deadline, the SA sorts them by their price, with the selected bid as the first element of the array with the highest price. Then the SA through a smart contract allows the buyer to make the transaction. The blockchain notifies via the event associated with the BA that wins the auction.

If the BA agrees to buy the energy, it audits the batteries of the selling entity; once the BA verifies that the entity has the energy it offers, it publishes the transaction via smart contract, and the SA finalizes the auction for that energy lot. If the BA refuses, then the SA selects the next bid from the array and repeats the process.

The BA continues to search for more offers if it has more bids or needs to buy more energy. The search stops when there is no more offer left or the energy demand is met. If the BA has more bids or needs more energy to buy, when searching offers, and they are more expensive than the offers of the MGO, the auction process is then carried out by the MGO instead of the SA.

Acknowledgments

This work has been supported by the Salamanca Ciudad de Cultura y Saberes Foundation under the Atracción del Talento program (CHROMOSOME project). Moreover, Amin Shokri Gaza-froudi acknowledges the support by the ministry of education of the Junta de Castilla y León and the European Social Fund through a grant from predoctoral recruitment of research personnel associated with the research project "Arquitectura multiagente para la gestión eficaz de redes de energía a través del uso de técnicas de intelligencia artificial" of the University of Salamanca. Also, Yeray Mezquita acknowledges the support by the predoctoral fellowship from the University of Salamanca and Banco Santander.

References

[1] A. Kumar, S. Grijalva, Graph theory and critical load-based distribution system restoration using optimal microgrids formation, in: 2018 Clemson University Power Systems Conference (PSC), September, IEEE, 2018, pp. 1–6.

[2] B.N. Stram, Key challenges to expanding renewable energy, Energy Policy 96 (2016) 728–734.

[3] F. Marra, G. Yang, Decentralized energy storage in residential feeders with photovoltaics, in: Energy Storage for Smart Grids-Planning and Operation for Renewable and Variable Energy Resources (VERs), Elsevier, 2015, pp. 277–294.

[4] N.L. Díaz, A.C. Luna, J.C. Vásquez, J.M. Guerrero, Equalization algorithm for distributed energy storage systems in islanded ac microgrids, in: IECON 2015-41st Annual Conference of the IEEE Industrial Electronics Society, November, IEEE, 2015, pp. 004661–004666.

[5] N.L. Diaz, A.C. Luna, J.C. Vasquez, J.M. Guerrero, Centralized control architecture for coordination of distributed renewable generation and energy storage in islanded ac microgrids, IEEE Trans. Power Electron. 32 (7) (2017) 5202–5213.

[6] A. Kahrobaeian, Y.A.R.I. Mohamed, Networked-based hybrid distributed power sharing and control for islanded microgrid systems, IEEE Trans. Power Electron. 30 (2) (2015) 603–617.

[7] D. Wu, F. Tang, T. Dragicevic, J.C. Vasquez, J.M. Guerrero, Autonomous active power control for islanded ac microgrids with photovoltaic generation and energy storage system, IEEE Trans. Energy Convers. 29 (4) (2014) 882–892.

[8] A. González-Briones, J.A. Castellanos-Garzón, Y. Mezquita Martín, J. Prieto, J.M. Corchado, A framework for knowledge discovery from wireless sensor networks in rural environments: a crop irrigation systems case study, Wireless Commun. Mobile Comput. 2018 (2018), https://doi.org/10.1155/2018/6089280.

[9] M. Francisco, Y. Mezquita, S. Revollar, P. Vega, J.F. De Paz, Multi-agent distributed model predictive control with fuzzy negotiation, Expert Syst. Appl. 129 (2019) 68–83, https://doi.org/10.1016/j.eswa.2019.03.056.

[10] V.-H. Bui, A. Hussain, H.-M. Kim, A multiagent-based hierarchical energy management strategy for multi-microgrids considering adjustable power and demand response, IEEE Trans. Smart Grid 9 (2) (2018) 1323–1333.

[11] H.K. Nunna, S. Doolla, Demand response in smart distribution system with multiple microgrids, IEEE Trans. Smart Grid 3 (4) (2012) 1641–1649.

[12] Y.F. Eddy, H.B. Gooi, S.X. Chen, Multi-agent system for distributed management of microgrids, IEEE Trans. Power Syst. 30 (1) (2015) 24–34.

[13] A. Goranović, M. Meisel, L. Fotiadis, S. Wilker, A. Treytl, T. Sauter, Blockchain applications in microgrids an overview of current projects and concepts, in: IECON 2017, 43rd Annual Conference of the IEEE Industrial Electronics Society, October, IEEE, 2017, pp. 6153–6158.

[14] pwr.company, pwr.company [online], Available: http://pwr.company.

[15] Power Ledger – Where Power meets Blockchain, Power Ledger - A New Decentralised Energy Marketplace [online], Available: https://powerledger.io.

[16] Transactive Grid, [online]. Available: http://lo3energy.com, May 2017.

[17] DAJIE!, [online]. Available: https://www.dajie.eu, May 2017.

[18] FAQs – easily charge your electric car with reliable charging station providers [online], Available: https://shareandcharge.com/en/faqs-2.

[19] NRGcoin, Smart Contract for green energy [online], Available: http://nrgcoin.org/faq.

[20] S. Nakamoto, Bitcoin: A Peer-to-Peer Electronic Cash System, 2008.

[21] V. Buterin, A next-generation smart contract and decentralized application platform, 2014. White paper.

[22] S. Omohundro, Cryptocurrencies, smart contracts, and artificial intelligence, AI Matters 1 (2) (2014) 19–21.

[23] I. Weber, X. Xu, R. Riveret, G. Governatori, A. Ponomarev, J. Mendling, Untrusted business process monitoring and execution using blockchain, in: International Conference on Business Process Management, 2016, pp. 329–347.

[24] Icertis, Smart contracts are transforming the way we do business, [online]. Available: https://www.icertis.com/resource/smart-contracts-are-transforming-the-way-we-do-business-featuring-gartner-research/, 2017.

Chapter 11

Blockchain-based protection schemes of DC microgrids

Navid Bayati[a], Amin Hajizadeh[b] and Mohsen Soltani[a]
[a]*Department of Energy Technology, Aalborg University, Aalborg, Denmark,* [b]*Department of Energy Technology, Aalborg University, Esbjerg, Denmark*

1. Introduction

Growing penetration of renewable energy sources (RESs), electric vehicles (EVs), energy storage systems (ESSs), and electronic loads in the modern power systems increases the use of LVDC microgrids [1]. Compared with AC micro-grids and regarding power flow, DC microgrids provide more efficiency between loads and RESs [2]. This is due to the lack of skin effect and reduction in conversion stages from the generation point to the DC loads. The power con-verters regulate the DC voltage, which enables the high stability in the DC microgrids to meet the strict power quality requirement of modern and critical electronic loads. In addition, in recent years, the efficiency of the converters has been improved to a comparable level with the AC power transformers, and it makes DC microgrids viable to use in the power systems [3].

On the other hand, to ensure a reliable, safe, and secure DC microgrid, consid-erable development, in the standardized protection scheme for DC microgrids, is essential. In addition, utilizing converters with output and input capacitive filters causes extremely high-amplitude transient currents during the fault in DC micro-grids. Therefore the high rise current during the fault increases the damaging prob-ability of converters, and it introduces that the faults in the DC microgrids must be isolated by a protection scheme with the lowest operation time [4].

Due to the increase in the reliability and resiliency of the DC microgrids, the best structure of this type of systems is the ring configuration [5]. Because of the bidirectional current flow in these systems, the traditional protection method would not work properly. In the DC microgrids with ring configuration, a communication-based method is one of the main protection schemes for the location and detection of faults. High-speed and communication-based differential protection techniques are suggested in different DC applications such as HVDC transmission [6], maritime [7], and aircraft [8]. However, the

Blockchain-based Smart Grids. https://doi.org/10.1016/B978-0-12-817862-1.00011-7

differential-based protection methods are not directly applicable in the LVDC microgrids. To design a high-speed differential protection scheme for these systems, different types of the high-speed protection equipment, such as synchronized communication links, processors, and circuit breakers (CBs), are necessary to install in the DC microgrid. Moreover an execution with a few samples may cause a misoperation in the stability and selectivity of the protection actions. In Sortomme et al. [9], differential relay method is considered as the main protection device (PD) of the system, and voltage-based protection method is implemented for the backup protection during the communication failure.

In other protection methods the communication between PDs is essential to update the value of the real-time current and voltage of the system to detect and isolate the fault currents. In addition, the RESs are monitored to make a decision about their status to disregard or include their fault contribution during the fault [10]. Consequently, using communication links in the DC microgrids is unavoidable to provide a real-time and adaptive protection method. The dependencies of physical and cyber components to each other in the DC microgrids excess the control methods for them. This problem is defined that in a close physical and cyber system (same as a DC microgrid with an adaptive protection system) minor malfunction and problem in the cyber system cause dangerous effects in the physical aspects of the system [11].

For solving this challenge, extensive researches have been studied to design data communication standards for DC microgrids. Affirming their criticality in the DC microgrids, requirements for communication of protection methods are presented in the IEC 61850 standard. This standard puts precise limitations on the signals of the communications in fault cases, such as generic object-oriented substation event signals, and 4-ms time restriction constrained on the sampled measured values [12]. Therefore proposing a protection technique based on the communication lines is a complex procedure. This problem can be more complicated during the use of a communication-based protection method that uses multiple agents and intelligent electronic devices (IEDs) to detect and isolate the fault in lower operation time.

It is an undeniable fact that the availability and robustness of the communication links are essential requirements to design an adaptive protection method. Consequently, solving the communication failure in DC microgrid protection schemes is one of the challenges. Despite the mentioned problems the penetration of communication links in the protection systems increases the vulnerability of these systems to the cyberattacks, which is reported as one of the newest challenges in the reliable and critical DC microgrids [13]. Consequently, this problem should be considered in the DC microgrid protection schemes, which increases the complexity of designing them. One of these problems is false data injection (FDI), which could cause bad data detection in the systems equipped by SCADA and make a huge error in state estimation of power systems [14]. Also, by causing FDI in a DC microgrid, it can remove some parts and components of the system and cause harmful commands and even a blackout in the system. In recent years, several researches are studied to protect the power systems against FDI attacks. For instance, Bobba et al. [15] suggested to secure some state variables and measurements to eliminate the FDI attacks in power systems. In addition, in Kosut et al. [16], the impact of FDI attacks on the electricity market operations is investigated by manipulating the real-time locational marginal price.

On the other hand, in the DC microgrid protection scheme, several components, such as smart measurements, switches, communication lines, and IEDs, are installed. These will provide data and trip signals for CBs during the fault to detect and isolate the fault. This high penetration of communication lines and processors into the DC microgrids increases the vulnerability of the system to cyberattacks. In this case, hackers can disrupt the switches or communication links to disable the protection of the system. In this case a backup protection scheme is essential, which considered the cyberattacks. In this case, several approaches are presented to defend and detect the cyberattacks based on the communication facilities [17].

However, the current communication links of measurements and input of IEDs in DC microgrids are not effective against the cyberattacks, even if the phasor measurement units (PMUs) are installed in the system. Due to their dependency on the global positioning system (GPS), these devices are sensitive to the cyberattack. Due to the distribution of the IEDs and measurement units in these systems, a backup scheme is essential to develop a security system for all effective components in the system. In this case the DC microgrid protection scheme can be considered as a distributed advanced measurement and communication infrastructure network, including knowledge storage, information monitoring, distributed data acquisition, and IEDs, on the demand and systems sides.

Nakamoto [18], for the first time, in 2008, proposes the blockchain concept to achieve the peer-to-peer directly electronic payments without using a third party, which is trusted. In blockchains, each peer plays a role as a node of the network, which all peers form a distributed network, and participates in the solution of a mathematical problem based on the hash to ensure the transaction integrities. All data are packed into a block and inserted in the existing blockchains. The recorded data of blocks are concertedly referred to a ledger [19]. Then, all blocks are synchronously updated to the network. Thus each peer retains the record of the same ledger.

In recent years the application of blockchain has been mostly limited to the financial issues, such as sustainable local energy markets [20], privacy in trading [21], and demand response programs [22]. However, due to the ability of blockchains in the security of the communication lines and IEDs, this new technology is a useful tool for designing a backup protection scheme to protect the DC microgrids against the cyberattacks.

Distinct from the aforementioned literature, in this chapter, a DC microgrid protection scheme based on the blockchain is proposed. Main contributions are as follows:

1. At the first stage the proposed protection scheme protects the DC microgrid against the fault current by using a differential protection method. The presented protection method uses communication lines between each PDs and IEDs. Therefore, during the fault, the faulty section is isolated by the proposed approach through the lowest operation time.
2. The proposed scheme improves the self-protection of modern DC microgrids against the cyberattacks by using blockchains. In the traditional power grids, the cyberattacks only can be performed by accessing to the measurement data and PDs locally, but, by increasing the penetration of the communication lines, hackers can control main components remotely. In the proposed technique the protection scheme uses a backup blockchain-based communication infrastructure to immune the protection scheme against cyberattacks.

This chapter is organized as follows: The general layout and components of the DC microgrids are described in Section 2. Section 3 defines the challenges in the protection of a DC microgrid. The concept of the blockchain and the application of this technology on the communication links are investigated in Section 4. In Section 5, different types of cyberattacks and the impact of them on the protection scheme are described. Section 6 proposes the differential-based protection method for DC microgrids as a primary PD. In this section the impact of fault resistances, measurement, and communication failure are investigated. In Section 7 the blockchain-based protection scheme for backup of the main protection systems is proposed. In this section the coordination between backup and primary PDs is evaluated to design a reliable DC microgrid. The results of the proposed method are presented for a case study and different scenarios to prove the effectiveness of this scheme in Section 8. In the last section the discussion and conclusion of this method are presented.

2. DC microgrids

DC microgrids are the most appropriate option for some specific power systems, such as rural power systems, aircraft, maritime, and office buildings. In addition, in these systems, the majority of loads and resources are based on the DC voltage. Today, telecommunication systems are used a DC voltage, typically 48 V, by using converters. In telecommunication systems a diesel generator is connected to the energy storage systems to support batteries during the power outages. Therefore it is reasonable to use DC voltage instead of the AC. In a DC microgrid the main components are loads, energy storage, converters, and resources. Moreover a small DC microgrid is shown in Fig. 1.

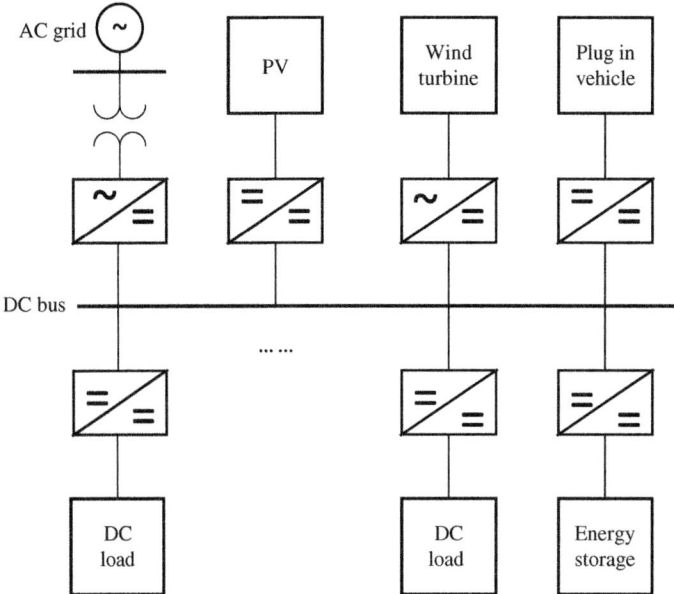

FIG. 1 Example of a small DC microgrid.

2.1 Power resources

Power resources in the DC microgrids have different types. Due to the DC voltage of these systems, fuel cells and photovoltaic (PV) systems, which generate the DC voltage, are appropriate sources for these systems by using a DC/DC converter. Other types of power resources, such as wind turbines (WTs) and microturbines, generate the AC voltage and should be connected to the DC bus by an AC/DC converter. The comparison of the resources that can be used in a DC microgrid is summarized in Table 1. Based on the application of the DC microgrid, suitable power resources should be selected.

Table 1 shows that based on the application of the DC microgrid, each power resources can be used, but, typically, in the standalone buildings, PV and WT provide efficient and fuel-less power. On the other hand, due to the variable output of these components, a power storage device is required for this renewable-based DC microgrids.

2.2 Energy storage system

The transient response and the uncertainty of some renewable energy resources such as PV and WT cause a need for them to combine with an energy storage system (ESS). Moreover an ESS provides an emergency power supply, load leveling, and power quality improvement [23]. A comparison between different technologies of ESSs is represented in Table 2 [24]. It is interesting to note that

TABLE 1 Main power resources technologies in DC microgrids.

Power resources	Voltage type	Efficiency (%)	Advantages	Disadvantages
WT	AC	50–80	• Power generation independent to the day or night • Low maintenance cost • Emission free • Small footprint • Low running cost	• Initial cost • Storage required • Wind fluctuates • Noise pollution • Visual pollution
PV	DC	40–45	• Clean and silent energy production • Small PVs can be used on unused spaces • It operates for a long period of time • Can be made for any size and powers	• Toxic chemical is used during the PV production • Expensive • Variable energy output • Cannot produce energy during night
Biomass	AC	60–75	• Widely available • Carbon neutral • Reduce the dependence to the fossil fuels • Cheaper than fossil fuel • Lower garbage	• Lower efficiency • Not entirely clean • Requires more space
Small hydro power	AC	90–98	• Clean • Controllable • Creates reservoirs • Useful to provide peak power	• Impact on the fish populations • Impact on water quality • Impacted by drought • Cannot be used in every places
Solar thermal	AC	50–75	• Low maintenance • Low operation cost	• Low energy density • Restricted scalability • Initial cost • Weather dependent

TABLE 1 Main power resources technologies in DC microgrids—cont'd

Power resources	Voltage type	Efficiency (%)	Advantages	Disadvantages
Fuel cell	DC	80–90	• Environmental friendly • Silent energy production • High energy efficiency • Energy flexibility	• Expensive infrastructure • Requires fuel • Requires temperature regulations
Microturbine	AC	80–85	• Small size • Low maintenance cost • Start up during the necessary times • Wide range of units size	• Requires fuel • Produces emission

TABLE 2 Main ESS technologies.

ESS	Efficiency (%)	Capacity (MW)	Lifetime (years)
Thermal energy storage	30–60	0–300	5–40
Lead acid battery	70–90	0–40	5–15
Nickel cadmium battery	60–65	0–40	10–20
Lithium-ion battery	85–90	0–1	5–15
Supercapacitors	60–65	0–0.3	More than 20

some technologies such as thermal energy storages have higher lifetime and capacities, but, because of the low capacity of LVDC microgrids and importance of the efficiency, the lead–acid and lithium-ion batteries are the best option for installation in these systems.

2.3 Converters

In a DC microgrid, both DC/DC and AC/DC converters are used to interconnect the AC and DC components to the main DC bus. These converters require to control the bidirectional power flow and the disturbances, such as voltage drop or faults. Typically the DC/DC converters are designed simpler than the AC/DC converters, which causes higher efficiency and lower cost [25]. In addition, during the fault, these converters can limit the value of the fault current, but, in the DC microgrid, due to the high rise of fault current and low tolerant of converters against fault current, the fault isolation time should be lower than AC microgrids.

3. Challenges in protection of DC microgrids

Besides the advantages of the DC microgrids, a protection system is required for the converter-based DC microgrids. Due to the lack of efficient standards for the protection of DC systems, focusing on the challenges of these systems is an important part of designing an efficient protection method. In this section the main challenges of DC microgrid protections are investigated.

3.1 Immunity against cyber attack

In recent years, due to the increase of the impact of the communication and the processor-based protection schemes, these systems are vulnerable against the cyberattack. Therefore designing a backup protection unit for the primary PDs to protect the system against cyberattacks is a vital challenge.

3.2 Low tolerant of converters

During the fault in the DC microgrids, the fault current reaches to the peak value in a short period. And, on the other hand, the tolerance of the converters to the high fault rises is low. Therefore a high fault current flows through the converters, which connect two main parts, for example, connecting the DC microgrid to the grid. Thus the fault current should be cleared in by a short operation time.

3.3 Unsuitability of the AC C.Bs

CBs interrupt the fault current in the cross zero point in the AC systems, but, in the DC systems, due to the absence of the cross zero point in the DC microgrids, the AC CBs cannot be implemented directly in the DC systems. Moreover, for preventing damage to the voltage source converters, fault in the DC microgrids should be interrupted in the less time compared with the AC systems [26]. Therefore power electronic devices such as integrated gate-commutated thyristor (IGCT) and insulated gate bipolar transistor (IGBT) are the best solution for clearing the fault.

3.4   Bidirectional fault current

Despite of the conventional power networks, the structure of the DC microgrids typically is ringlike. Thus the fault current of these systems is bidirectional, and the protection scheme of radial system cannot be implemented in DC microgrids. In addition, due to the variation of the topology and uncertainty of renewable energy resources, the short-circuit level in the DC microgrids is not constant.

3.5   Unsuitability of the AC protection methods

In recent years, several researches are proposed for the protection of AC microgrids, which can protect the AC systems against the fault situations. However, the majority of these approaches use the frequency and phase of the fault current. But, due to the absence of these parameters in a DC microgrid, most of the AC protection methods cannot be implemented in the DC systems.

4.   Cyber attacks

In this section, different types of the cyberattacks and the impact of them on the communication link and protection system are investigated. Cyberattacks on the communication links interrupt signals of relays and measurement units and consequently interrupt in the clearing of the fault by CBs. The cyberattacks are divided into two different types, SV and GOOSE messages and network security attacks, as shown in Fig. 2 [27].

4.1   SV and GOOSE message cyber-attacks

The SV and GOOSE messages are the two communication protocols of the IEC 61850. The SV messages send the measured values of the sensors or

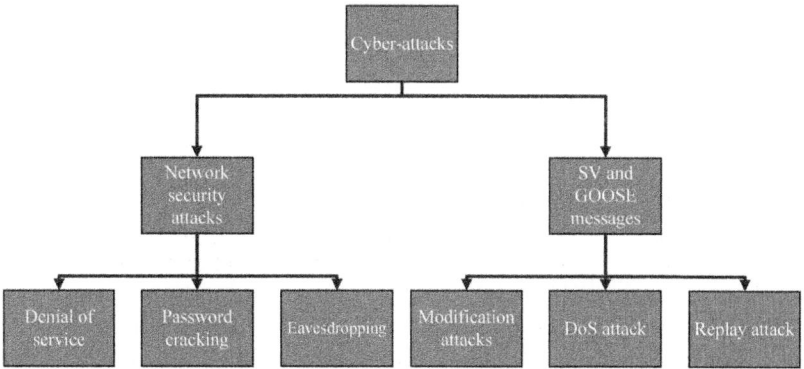

FIG. 2   Cyberattack types.

measurement units to the PDs, and the GOOSE messages send the trip signal to the CBs to clear the fault location. Based on the standards, these signals should be sent less than 4 ms. In the protection systems, both messages use Ethernet network, for short lengths, and phasor measurement unit (PMU), for long lengths.

1. Replay attacks: In this cyberattack the SV packet, including measured values, is captured by attacker and then sends it to another PD. For GOOSE messages, attacker captured this signal and sends a trip signal to the CB during the normal operation mode.
2. Denial-of-service (DoS) attacks: DoS is an attack that causes users not to access to the service. In this type of attack, relays are prevented to respond to the correct messages from other measurement units, PDs or relays. Attacker sends a large number of the messages and data to the targeted relay; then the target cannot respond to the correct signals. Another way of DoS attack is GOOSE poisoning attack. In the poisoning attack, GOOSE messages with a higher sequence number than those sent by the relay are accepted by CBs. Therefore all trip signals from the relay are considered as an invalid signal, and only the incorrect signals from the attacker will be accepted. GOOSE poisoning attacks include semantic, high-rate flooding, and high-status number attacks.
   - Semantic attacks: The status number in the GOOSE message is fixed, and the status change rate is determined by the attack.
   - High-rate flooding attacks: A range of fake GOOSE messages are multicasted with a higher number of status. Then a status number more than the expected status number of CB are employed by the fake GOOSE message.
   - High-status number attacks: A single fake GOOSE message with a high-status number is sent to the CB [28].
3. Modification attack: The message between the relay and CBs are changed without permission of the relay. The GOOSE messages are captured and modified by attacker to another message, and then, CBs can be controlled by attacker. Also, attacker can send a false analog signal for SV packets to a controller in the DC microgrid to cause an artificial fault or outages. Another tool that attackers can use for modification attacks is malware script [29]. This type of attack is local; hence, it requires to be installed inside of the network. This method captures the exchanged messages between relays and injects GOOSE messages to the IEC 61850 network.

4.2 Network security cyber-attacks

The target of this type of attack is communication links to access and change data [30]. Different types of cyberattacks are eavesdropping attacks, password cracking attempts, and DoS.

1. Eavesdropping attacks: This type of attacks is a local type and requires to use the local network to steal packets that are exchanged in the protection system. One type of cyberattack is the address resolution protocol (ARP) cache poisoning. Attacker converts the IP address into the false MAC address using ARP communication protocol, and it enables the attacker to capture all packets. Another type of eavesdropping attacks is switch port stealing in which false signals are transmitted to the CB MAC address and it allowed the attacker to connect to the system.

2. Password cracking attempts: All devices such as relay or other PDs required a password to get access to the system; therefore attackers use password cracking attempts and guess the password to gain access to the protection system [31]. Thus, by accessing to the protection system, attackers can send false signals to the CBs and isolate the healthy parts.

3. DoS: DoS effects on the communication links between PDs, and attackers transmit a continuous fake synchronization signal to the relay to disturb the connection between the operator and the relays [32]. The DoS attack is run by using different protocols such as Telnet, HTTP, and FTP on the relays at the same time.

5.   Blockchain structure

A blockchain is a chain of a number of blocks that records the value and amount of data. The structure of blockchain is depicted in Fig. 3. Each block is connected to the previous block through references. In this stage the hash value of the previous block and that of the first block are called *parent* and *genesis* block, respectively.

A block includes block body and block header, as shown in Fig. 3, and the block header includes the following:

- nonce: a 4-byte field,
- nBits: compact format of the current hash,
- timestamp: current timestamp,
- Merkle tree root hash: the hash value of all transactions,
- parent block hash: the hash value of the current point the previous block,
- block version: determines that set of block validation rules to follow.

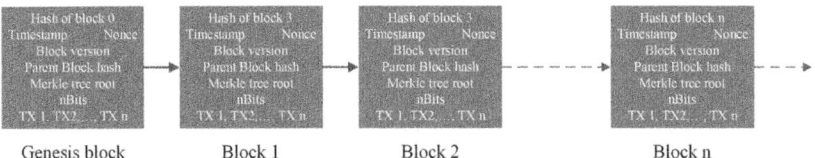

| Genesis block | Block 1 | Block 2 | Block n |

FIG. 3 A simple structure of the blockchain.

A block body is made by transactions and a transaction counter. The size of each transaction and the block size determine the maximum number of transactions that a block can take.

Moreover, each user accesses to public and private keys. The transactions are signed by the private key, and the signed transaction is available for everyone in the network by using a public key. A digital signature consists of verification and signing phase. For instance, first, a hash value is generated from the hash values, when a user wants to sign transactions. Then, user codes these hash values by using the private key and sends it to other users with the original data. The second user verifies the transactions by comparing the decoded hash and the hash value derived from the received data. Elliptic curve digital signature algorithm (ECDSA) is the digital signature method that is used in the blockchain.

The characteristics of blockchains are summarized as follows:

- Auditability: When a transaction is validated and recorded on the blockchain by timestamp, user verifies and traces the previous records via access to all nodes of a distributed network.
- Anonymity: By using generated addresses, users can interact with the blockchain network. Moreover, users can make several addresses to prevent being identified. This algorithm causes appropriate privacy on the blockchain transactions.
- Persistency: In the distributed network, it is almost impossible to tamper, because each transaction spreads into the network require confirmation and recordation in distributed blocks. In addition, each transaction will be checked and spread block validated by nodes. Thus any false data can be detected.
- Decentralization: In the traditional centralized transaction systems, transactions require to be validated by a central unit. On the other hand a peer-to-peer transaction can be conducted for a transaction in the blockchain system without validating by a central unit. Therefore blockchain significantly reduces server costs.

6.   Proposed blockchain-based DC microgrid protection technique

In this chapter a novel blockchain-based differential protection method for DC microgrids is proposed. Instead of shutting down the protection system during the cyberattacks, the proposed technique detects the fault by a blockchain-based backup system. In the first stage the proposed differential protection system detects the internal fault and ignores the external faults, by a high-speed communication line and relay. However, during the cyberattacks, this method is connected to a blockchain system as a backup of the communication system that prevents system against the cyberattack and communication link failure.

6.1 Differential fault detection

A protection system should distinguish between internal and external faults and only isolate the internal faults. During the fault the direction of the fault current will be to the fault location; therefore, by using the direction of two sides of the differential protection system, the type of fault can be detected. The implementation of the protection method and different location of faults are shown in Fig. 4.

In this system the differential relay receives the value of line currents from two measurements at two ends of the line. During the fault a variable for each side of relay is defined, S_1 and S_2 for bus 1 and 2, respectively; and for clockwise direction the value of S_1 and S_2 will be 1; and for counterclockwise direction the value of S_1 and S_2 will be -1. Therefore, during the fault, the internal fault can be detected by

$$\begin{cases} S_1 = S_2 \to \text{External fault} \\ S_1 = -S_2 \to \text{Internal fault} \end{cases} \tag{1}$$

$$S_T = \frac{|S_1 S_2 - 1|}{2} \tag{2}$$

where S_T is 0 for external faults and 1 for internal faults.

During the fault, the fault current transient is divided into two stages. The first stage is capacitor discharge, which capacitors start to discharge through the faulty path. This stage starts from the first moments of the fault and makes a high peak, as shown in Fig. 5. The second stage is the freewheeling diode operation, and it will start after dropping the voltage to zero. This stage shows the reversion of voltage of converter, and diode of converter starts to conduct.

The capacitors are the equivalent of converter capacitors at both ends of the lines. During the first moments of the fault, a high rise fault current is injected from these capacitors to the fault location.

In a bidirectional DC microgrid, during the fault, a line can be modeled by Fig. 6. In this system the differential value of the derivative of fault current for both sides of the line can be calculated by

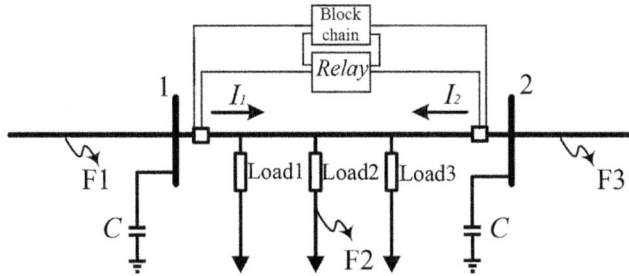

FIG. 4 Implementation of the proposed differential protection system.

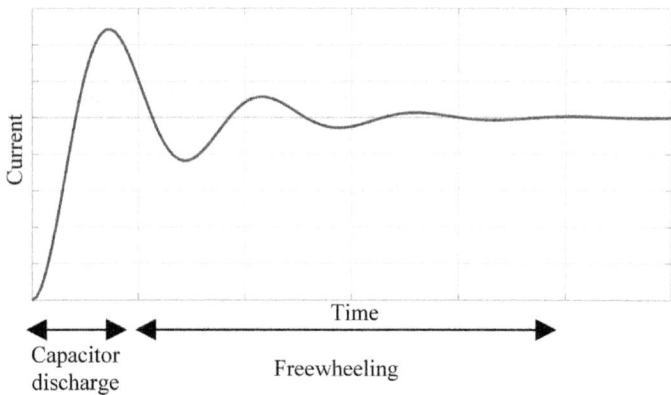

FIG. 5 Capacitor and freewheeling stages.

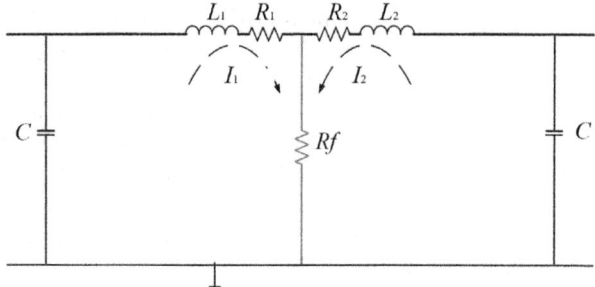

FIG. 6 The equivalent model of a line during the fault.

$$\alpha = \frac{di_1}{dt} - \frac{di_2}{dt} = \frac{L_1 d^2 i_1}{R_f \ dt} + \frac{R_1 + 2R_f \ di_1}{R_f \ dt} + \frac{1}{CR_f} i_1 \tag{3}$$

where i_1 and i_2 is the measured current of bus 1 and 2, respectively. L_1 and R_1 is the value of inductance and resistance from bus 1 to the fault location. R_f is the fault resistance, and C is the capacitance of the converters. Therefore if the value of α change to a value higher than a threshold, the differential relay sends the trip signal to the CBs. The value of the threshold can be obtained for the minimum value of fault current. For this situation, first, the maximum value of fault resistance should be estimated. Based on AC protection standards, in overload situations, a load can be fed with 25% more power than the normal conditions. Thus the maximum overload current is $1.25I_{load}$, which I_{load} is the nominal power of loads, and the current more than this value should be considered as a fault current. Therefore the maximum value of fault resistance is calculated by

$$R_f = \frac{4V_n}{I_{load}} \tag{4}$$

where V_n is the nominal voltage of the system. The minimum value of α or threshold is obtained by

$$\alpha_{min} = 2\frac{di_1}{dt} + \frac{I_{load}}{4CV_n}i_1 \tag{5}$$

Consequently the trip signal of the relay is made as follows:

$$\alpha_{min} - S_T\alpha \leq 0 \rightarrow trip \tag{6}$$

6.2 Blockchain backup protection system

In this chapter a new application of blockchains is proposed to manage the protection system during cyberattacks and communication failure. During the conditions without cyberattack or communication failure, the signal to the relay from the measurement unit and the signal from the blockchain to the relay is the same. However, during the cyberattack to the main communication link, the encrypted value from the blockchain will be different. Then the relay detects the cyberattack and only considers the data from the blockchain. Therefore the values of i_1 and i_2 in Eqs. (1)–(6) will change to the decoded value of current from blockchain.

However, during the cyberattack, by using the blockchain system, a small delay will be added to the protection system. Moreover, by installing another measurement unit at each bus, the reliability of the protection system against the cyberattack will increase, but the cost also will increase.

7. Simulation results

A MATLAB simulation has been performed for a line of a DC microgrid as shown in Fig. 4. Two sides of the understudy line are connected to a part of DC microgrid with power resources; therefore the fault current is bidirectional. The proposed method is tested for different cases such a HIF and cyberattack to the communication link.

7.1 Fault detection and isolation without cyber-attack

An interval line to ground fault with fault resistance 1 Ohm is occurred, at the 33% of the line from bus 1. The fault current seen from bus 1 is shown in Fig. 7. Then the proposed differential relay detects and isolates the fault in 1.4 ms, as depicted in Fig. 8.

As shown in Fig. 8, during the first stage of the fault, the discharge of the capacitor causes a high rise current through the line. The peak of fault current,

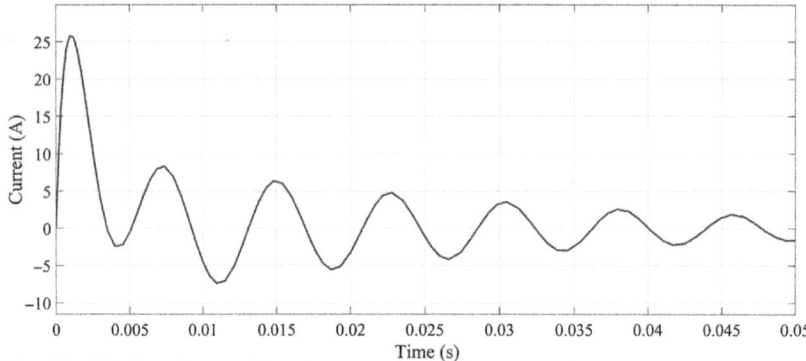

FIG. 7 Fault current without using an isolation method.

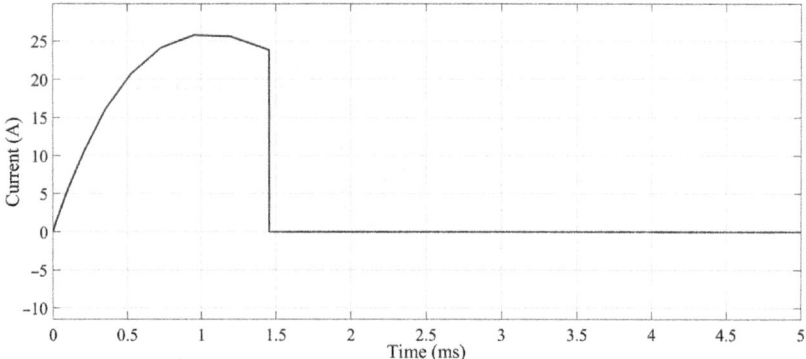

FIG. 8 Detecting and isolating the fault with fault resistance 1 Ohm.

in this case, is almost 25 A, and therefore the differential relay detects the fault at 1.4 ms and sends the trip signal to CBs of both sides of the line. In other situation a HIF with fault resistance of 10 Ohms has occurred at the middle of the line, as shown in Fig. 9. The fault current is reached to approximately 4 A, and then the fault is detected and cleared in 0.5 ms.

7.2 Fault detection and isolation by considering a cyberattack

As mentioned before, one of the threats to a protection scheme is cyberattack; therefore a blockchain system is linked to the differential relay to protect the system against cyberattack. As shown in Fig. 10, a cyberattack has occurred on the communication link of the differential relay. Therefore the blockchain system detects the cyberattack and sends the encrypted data to the differential relay. Consequently, the waveform of fault current before by implementing the protection system.

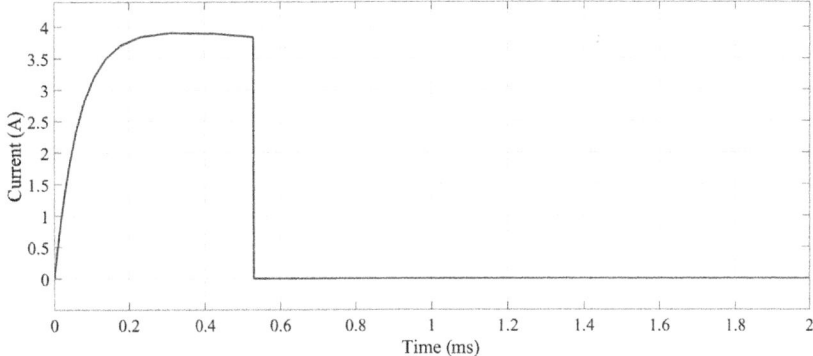

FIG. 9   Detecting and isolating the fault with fault resistance 10 Ohms.

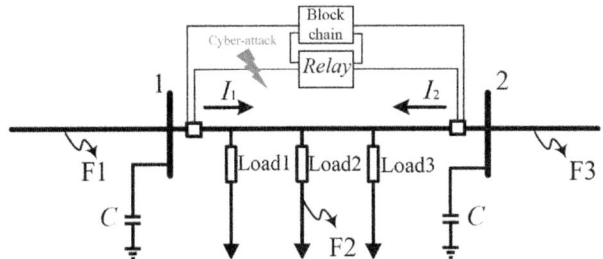

FIG. 10   A cyberattack on the protection system.

During the cyberattack, if a fault with fault resistance 1 Ohm has occurred in the F_2, the differential relay detects and isolates the fault in less than 29 ms. In this case, due to the encrypting by blockchain and detecting cyberattack, a delay is observed in the operation of the protection scheme, as shown in Fig. 11. Also a HIF is occurred at the middle of the protected line by fault resistance 10 Ohms, which the isolated fault current is shown in Fig. 12.

The results show the proposed protection method, detect and isolate the fault in several milliseconds, and by using a blockchain system cyberattack cannot affect the protection system.

8.   Conclusion

This chapter has proposed a blockchain-based differential protection technique for DC microgrids. The proposed protection system consists of a differential relay connected to a blockchain system for protecting the system against cyber-attacks. The differential relay detects the internal faults and isolates the faulty line before ending the capacitor discharge state. Moreover a new threshold selection approach is proposed, which able the differential relay to detect the HIFs. The proposed method is tested and validated by simulation.

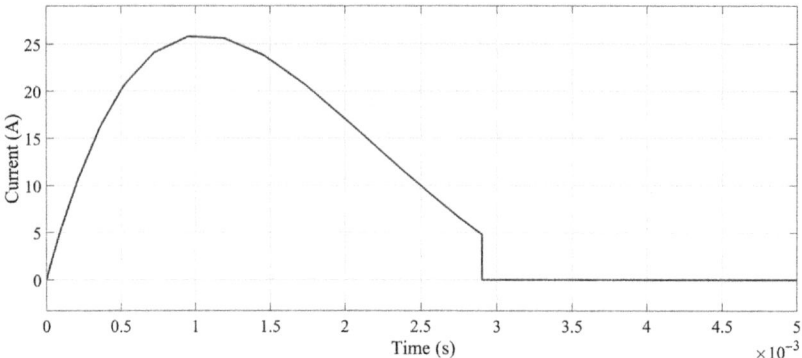

FIG. 11 Detecting and isolating the fault with fault resistance 1 Ohm during cyberattack.

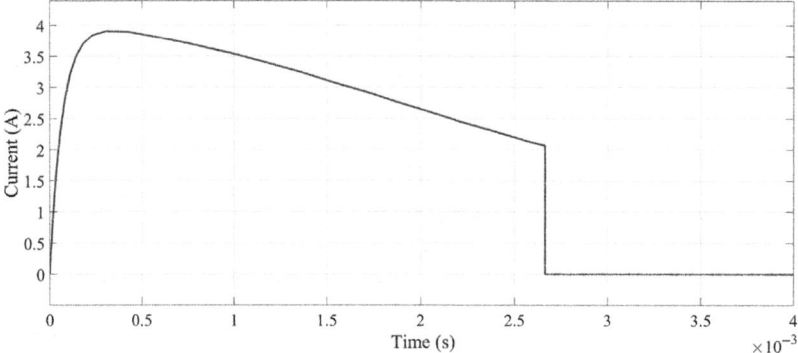

FIG. 12 Detecting and isolating the fault with fault resistance 10 Ohms during cyberattack.

The simulation results show that the proposed protection scheme detects and isolates the HIFs by considering the different situations for cyberattacks and also the backup blockchain system increases the reliability of the protection system by decreasing the impact of communication failure. During the fault the HIF is detected in less than 0.5 ms, and by considering a cyberattack the HIF is also detected in less than 2.5 ms. The results show the effectiveness of the proposed method against cyberattacks.

References

[1] N. Bayati, A. Hajizadeh, M. Soltani, Protection in DC microgrids: a comparative review, IET Smart Grid 1 (3) (2018) 66–75.

[2] N. Bayati, A. Hajizadeh, M. Soltani, Impact of faults and protection methods on DC microgrids operation, in: 2018 IEEE International Conference on Environment and Electrical Engineering and 2018 IEEE Industrial and Commercial Power Systems Europe (EEEIC/I&CPS Europe), IEEE, 2018, pp. 1–6.

[3] H. Ergun, D.V. Hertem, Comparison of HVDC and HVDC technologies, in: HVDC Grid: For Offshore and Supergrid of the Future, Wiley-IEEE Press, 2016, Chapter 4. pp. 79–96.

[4] C. Li, P. Rakhra, P. Norman, P. Niewczas, G. Burt, P. Clarkson, Modulated low fault-energy protection scheme for DC smart grids, IEEE Trans. Smart Grid 11 (2019) 84–94.

[5] N. Bayati, A. Hajizadeh, M.N. Soltani, Localized fault protection in the DC microgrids with ring configuration, in: 28th International Symposium on Industrial Electronics (ISIE), IEEE Press, 2019.

[6] D. Tzelepis, A. Dysko, G. Fusiek, J. Nelson, P. Niewczas, D. Vozikis, P. Orr, N. Gordon, C. Booth, Single-ended differential protection in MTDC network using optical sensors, IEEE Trans. Power Del. 32 (3) (2017) 1605–1615.

[7] K. Satpathi, A. Ukil, J. Pou, Short-circuit fault management in DC electric ship propulsion system: protection requirements, review of existing technologies and future research trends, IEEE Trans. Transport. Electrific. 4 (1) (2018) 272–291.

[8] S. Fletcher, Protection of physically compact multi-terminal DC power system, Ph.D. thesis, Dept. Elect. Eng., Strathclyde Univ., Glasgow, UK, 2013, pp. 54–55.

[9] E. Sortomme, S.S. Venkata, J. Mitra, Microgrid protection using communication-assisted digital relays, IEEE Trans. Power Del. 25 (4) (2010) 2789–2796.

[10] H. Wan, K.K. Li, K.P. Wong, An adaptive multiagent approach to protection relay coordination with distributed generators in industrial power distribution system, IEEE Trans. Ind. Appl. 46 (5) (2010) 2118–2124.

[11] R. Akella, H. Tang, M.M. Bruce, Analysis of information flowsecurity in cyber-physical system, Int. J. Crit. Infrastruct. Protect. 3 (2010) 157–173.

[12] H.F. Habib, C.R. Lashway, O.A. Mohammed, A review of communication failure impacts on adaptive microgrid protection schemes and the use of energy storage as a contingency, IEEE Trans. Ind. Appl. 54 (2) (2017) 1194–1207.

[13] Y. Mo, et al., Cyber–physical security of a smart grid infrastructure, Proc. IEEE 100 (1) (2012) 195–209.

[14] Y. Liu, P. Ning, M.K. Reiter, False data injection attacks against state estimation in electric power grids, Proc. ACM Comput. Commun. Soc. (CCS) (2009) 21–32.

[15] R.B. Bobba, K.M. Rogers, Q. Wang, H. Khurana, K. Nahrstedt, T.J. Overbye, Detecting false data injection attacks on DC state estimation, in: Proc. Preprints 1st Workshop Secure Control Syst. (CPSWEEK), 2010, pp. 1–9.

[16] O. Kosut, L. Jia, R.J. Thomas, L. Tong, Malicious data attacks on the smart grid, IEEE Trans. Smart Grid 2 (4) (Dec. 2011) 645–658.

[17] Y. He, G.J. Mendis, J. Wei, Real-time detection of false data injection attacks in smart grid: a deep learning-based intelligent mechanism, IEEE Trans. Smart Grid 8 (5) (Sep. 2017) 2505–2516.

[18] S. Nakamoto, Bitcoin: A Peer-to-Peer Electronic Cash System [Online], Available: https://bitcoin.org/bitcoin.pdf, 2008.

[19] G. Liang, S.R. Weller, F. Luo, J. Zhao, Z.Y. Dong, Distributed blockchain-based data protection framework for modern power systems against cyber attacks, IEEE Trans. Smart Grid 10 (3) (2018) 3162–3173.

[20] E. Mengelkamp, B. Notheisen, C. Beer, D. Dauer, C. Weinhardt, A blockchain-based smart grid: towards sustainable local energy markets, Comput. Sci. Res. Dev. 33 (1–2) (2018) 207–214.

[21] N.Z. Aitzhan, D. Svetinovic, Security and privacy in decentralized energy trading through multi-signatures, blockchain and anonymous messaging streams, IEEE Trans. Depend. Secure Comput. 15 (5) (2016) 840–852.

[22] C. Pop, T. Cioara, M. Antal, I. Anghel, I. Salomie, M. Bertoncini, Blockchain based decentralized management of demand response programs in smart energy grids, Sensors 18 (1) (2018) 162.

[23] J. Barton, D. Infield, Energy storage and its use with intermittent renewable energy, IEEE Trans. Energy Convers. 19 (2) (2004) 441–448.

[24] E. Planas, J. Andreu, J.I. Gárate, I. Martínez de Alegría, E. Ibarra, AC and DC technology in microgrids: a review, Renew. Sustain. Energy Rev. 43 (2015) 726–749.

[25] N. Mohan, T. Undeland, W. Robbins, Power Electronics: Converters, Application and Design, second ed., Wiley, New York, 1995.

[26] N. Bayati, A. Hajizadeh, M. Soltani, Accurate modeling of DC microgrid for fault and protection studies, in: 2018 International Conference on Smart Energy Systems and Technologies (SEST), IEEE, 2018, pp. 1–6.

[27] H.F. Habib, C.R. Lashway, O.A. Mohammed, A review of communication failure impacts on adaptive microgrid protection schemes and the use of energy storage as a contingency, IEEE Trans. Ind. Appl. 54 (2) (2017) 1194–1207.

[28] N. Kush, E. Ahmed, M. Branagan, E. Foo, Poisoned GOOSE: exploiting the GOOSE protocol, in: Proceedings of 12th Australasian Information Security Conference, vol. 149, 2014, pp. 17–22.

[29] M.T.A. Rashid, S. Yussof, Y. Yusoff, R. Ismail, A review of security attacks on IEC61850 substation automation system network, in: Proceedings of 6th International Conference on Information Technology and Multimedia, Putrajaya, Malaysia, 2014, pp. 5–10.

[30] US-CERT, Understanding Denial-of-Service Attacks, [Online]. February 06. Available: https://www.us-cert.gov/ncas/tips/ST04-015, 2013. (Accessed 8 March 2017).

[31] Oxid.it, Brute-Force Password Cracker [Online], Available: http://www.oxid.it/ca_um/topics/brute-force_password_cracker.htm. (Accessed 8 March 2017).

[32] K. Choi, X. Chen, S. Li, M. Kim, K. Chae, J.C. Na, Intrusion detection of NSM based DoS attacks using data mining in smart grid, Energies 5 (2012) 4091–4109.

Index

Note: Page numbers followed by *f* indicate figures, *t* indicate tables, and *b* indicate boxes.